ALLE · ZEIT · WACH · 1842

aspects, and updating these research findings on individual crops in the form of a volume is of value to the flower grower.

Although postharvest shrinkage in almost all flower crops is caused and/or accelerated by ethylene, plant response to ethylene varies with the species. Subjective evaluation of the pertinent research developments on the process of senescence as applied to different ornamental plants must benefit all those who handle these crops.

In short, consumer satisfaction, an end result of the improved technologies in this industry, can be increased, and this volume can go a long way towards achieving that goal.

October 1989

D. K. Salunkhe
N. R. Bhat
B. B. Desai

Preface

The floricultural industry has been undergoing an unprecedented revolution in terms of the type of commodity produced and the production and marketing technology in both developed and developing countries. As a result of this revolution, as we know today, there is a flower for every purpose and for every person in the world, as is evident from the slogan of the Society for American Florists: "say it with flowers". In recent years, the Latin American and European countries have become sizeable competitors for the North American fresh flower markets and the trend continues growing.

Like any other crop production, floricultural production can be divided into three basic factors: (1) production costs (2) quality (3) transportation costs. All these must be optimum for this area or industry to be safe from competition. With increasing consumer awareness and the current recession, the pressure from the artificial floral products industry and also of neighbouring countries on the American fresh flower industry, and continued competition even amongst the growers, wholesalers and retailers, quality in floricultural industry is becoming increasingly important to all those concerned with handling these products. The visual quality aspects of the product are the sole determiner of consumer acceptability in this industry and, unlike fruits and vegetables, flowers cannot be marketed by just discarding the damaged portion.

Like other perishable products, such as fruits and vegetables, the postharvest shrinkage in the flower industry is very high (20% – 50%) and in certain developing countries can even be as high as 100%. These losses become especially important when dealing with the export of fresh flowers to distant and foreign markets. Factors such as higher moisture contents, susceptibility of flower crops to postharvest infections and storage conditions make them more vulnerable during postharvest handling. The sale of quality fresh cut flowers and other ornamental crops can be increased at least to the extent of this shrinkage, if improved biotechnologies for loss reduction are followed.

The nature and extent of postharvest damage is typical for each crop or cultivar. The preharvest growing conditions such as growing media, nutrition, light, temperature and relative humidity and the use of growth-regulating chemicals influence the postharvest qualities, and the plant response to these preharvest growing conditions varies with the crop. There has been a considerable amount of research done on these

Prof. Dr. DATTAJIRAO K. SALUNKHE
Dr. NARAYANA R. BHAT
Dr. BABASAHEB B. DESAI

Mahatma Phule Agriculture University
Rahuri District
Ahmednagar (M. S.)
PIN 413722
India

ISBN-13: 978-3-642-73805-0 e-ISBN-13: 978-3-642-73803-6
DOI: 10.1007/978-3-642-73803-6

Library of Congress Cataloging-in-Publication Data. Salunkhe, D.K. Postharvest biotechnology of flowers and ornamental plants / D.K. Salunkhe, N.R. Bhat, B.B. Desai. p. cm. Includes bibliographical references. 1. Cut flowers – Postharvest technology. 2. Plants, Potted – Postharvest technology. I. Bhat, N.R. (Narayana R.), 1953– . II. Desai, B.B., 1941– . III. Title. SB442.5.S25 1989 645.9'66 – dc 20 89-21592

Softcover reprint of the hardcover 1st edition 1990

Typesetting: K+V Fotosatz GmbH, Beerfelden

2131/3145-543210 – Printed on acid-free paper

D. K. Salunkhe · N. R. Bhat · B. B. Desai

Postharvest Biotechnology of Flowers and Ornamental Plants

With 14 Figures

Springer-Verlag Berlin Heidelberg New York
London Paris Tokyo Hong Kong

Contents

Chapter 1 Introduction

Flowers possess the great gift of establishing a tie with nature, thus satisfying an innate human need which is becoming increasingly evident in every society. As a result of continuous evolution, the floricultural industry, as we know it today, apart from being an art, is a highly specialized profession involving several basic scientific principles. It is now to offer floral products to suit every occasion and every kind of person in the world.

The flower is a unique organ when compared with seeds, fruits and vegetables. Unlike seed or fruit, which are single morphological units, a flower or inflorescence is composed of many morphological units such as sepals, petals, androecium, gynoecium, stem and often leaves. Each of these morphologically and physiologically complex units interacts with the others, making the flower more complex than the other plant organs [1]. The interaction between flowers, stems and leaves (e.g. movement of sugars from leaves to flower) influences the water balance and postharvest quality of flowers [2, 5–10]. Special care is thus needed in the development and postharvest handling of cut flowers. Postharvest life of cut flowers depends upon several pre- and postharvest factors, nutritional status of the growth medium, areal and root temperature, photoperiod and light intensity during the growth period in the greenhouse, conditions during harvesting, grading, packaging, transportation, and composition of "pulsing" and "bud opening" solutions. According to Halevy and Mayak [1], the postharvest handling techniques of fruits and vegetables, many of which are harvested after complete development, are directed primarily at delaying senescence and at maintaining the harvested product in "fresh" condition for as long as possible. In contrast to this, two distinct physiological phases have been identified in flowers: (a) bud growth and development of flower to full opening, and (b) maturation, senescence and wilting. These authors further pointed out that the extension of vase life of cut flowers thus, involves co-ordination of two seemingly conflicting processes: the promotion of growth during the first phase, and retardation of senescent processess during the second phase.

1.1 Production, Utilization and the International Trade

1.1.1 Production and Utilization

Like fruits and vegetables, flowers and ornamental plants suffer from inadequate production statistics. Several floral crops are often grouped together when estimating production and sales.

Table 1.1. Export of flowers and ornamental plants from The Netherlands to Europe during 1980

Export source	Export value	
	($ 1000)	(%)
Belgium	25765	2.7
Denmark	11885	1.3
West Germany	643756	68.6
Great Britain	34748	3.7
France	101928	10.9
Italy	18835	2.0
Norway	3982	0.4
Austria	31199	3.3
Spain	4583	0.5
Sweden	25990	2.8
Switzerland	36017	3.8
Total Europe	938688	100.0

An unpublished report by the Aalsmeer Auction (1981) [3].

The Netherlands is the major source of floral crops, exporting a wide variety of floral products, foliage plants and other ornamentals to northern Europe, North America, the Far East and Third World countries. According to Staby and Robertson [3], the major volume goes to Europe, accounting for about $ 1 billion in 1980 (Table 1.1). The Netherlands exported another $ 31 million in flowers to Colombia, Israel and other countries.

Although the USA is a major producer of flowers and ornamentals, it is also a net importer of cut flowers and greens, along with most other countries in Europe (Table 1.2). About 55% of the production of The Netherlands is exported. The USA is still considered as a good export market by producers in The Netherlands, Colombia and Israel [3].

1.1.2 International Flower Transport Trade

In tropical countries, flowers are grown outdors (or in plastics houses, saran houses, lath houses). Owing to the abundance of outdoor flowers, there is a limited demand for flowers on the local market in these countries, so that over 90% of the flowers produced there is exported. In contrast, in countries with more temperate climates, floriculture is a highly specialized profession, and as such the public has to depend upon flowers produced in a protected environment. There is also great demand for imported flowers, greens and other ornamentals in these countries.

During the early and mid-20th century in some of the Western countries, especially in The Netherlands, USA, West Germany, France and Italy, there has been a tremendous expansion in the area under greenhouses producing flowers and ornamentals [11]. The area under greenhouses in the USA, for example, increased from 14.85 million m^2 to 17.72 million m^2 (about 19% in-

Table 1.2. Estimates of domestic production and consumption of flowers and ornamental plants in Europe and North America in 1977, 1982 and 1987

Country	Domestic value ($ 1000)					
	1977		1982		1987	
	Domestic production	Domestic production	Domestic production	Domestic production	Domestic production	Domestic production
West Germany	1280	1975	1690	2795	2085	3600
France	810	860	1175	1335	1475	1720
Italy	800	745	1100	945	1320	1150
The Netherlands	1015	460	1620	705	2135	905
Switzerland	140	210	180	300	225	395
Sweden	120	145	150	270	195	355
Austria	140	180	180	260	225	340
Great Britain	120	145	160	360	195	580
Belgium	170	130	230	210	285	290
Norway	105	120	140	195	170	260
Denmark	145	85	210	120	260	155
Finland	55	60	70	90	90	115
Spain, Greece and Portugal	110	100	200	200	400	340
USA	1800	2000	2200	2600	305	3800
Canada	130	180	165	195	245	285
Total ($)	6940	7740	9470	10355	12355	14290

An unpublished report by the Aalsmeer Auction (1981) [3].

crease) during the period from 1949 to 1959. There was even greater expansion in outdoor flower production, with the result that the value of the products of floriculture increased by approximately 53% from 1949 to 1959. This trend continued until the recession in the early mid 1970's. According to the 1981 census, the wholesale value of the floricultural products in the USA was $ 1.02 million, which was 7% higher than that of 1980. A similar trend was evident in some of the European and tropical countries. According to the 1977 census, the area under greenhouses in The Netherlands was estimated to be about 3400 ha, which was about double the size of the industry in 1970. The wholesale value in the floricultural industry in 1977 was three times that of 1970, and accounted for 35% of the total horticultural industry in The Netherlands. Some of the tropical countries, such as Israel, Japan, Kenya, Mexico, Columbia and other Latin American countries, also showed similar expansion. Floriculture developed even faster in these countries than in some of the development countries.

This unprecedented revolution in floriculture resulted in the establishment of major flower production centres such as the northern countries of Europe, North America (Canada and the United States) and Japan. Comparatively limited quantities of flowers are produced in the East European coutries, Russia and Australia. In recent years, some of the tropical countries, in Central and South America, like Colombia or South Africa (Kenya), Israel and India have become leading production centres and have emerged as sizeable competitors for the North American and European markets for fresh flowers and ornamentals [11].

1.1.3 Nature of the Floriculture Industry

The Netherlands. The present floricultural industry of The Netherlands is the result of years of evolution from a nearly non-existent flower trade, which is believed to have originated in the 1600's [5]. One of the unique characteristics of the Dutch flower industry is its dependence on glass as a major covering material for greenhouses, which accounts for nearly 95% of the total area under protected cultivation. The majority of the Dutch flowers are produced in the western part (in and around Alsmeer). Another unique feature of the Dutch flower industry is its marketing system, which is the most efficient in the world. Due to the highly specialized selling system, the Dutch flower grower is able to specialize in one or more crops, and is thus able to produce better quality flowers and ornamental plants. Forty percent of floricultural products are sold to retail and only 15% to mass market outlets [11].

About 10–15 years ago, many vegetable growers switched over the flower production in this country; however, this trend seems to have stopped at present. During recent years, increasing heating costs have forced many medium-sized cut flower operations to change to pot plant production, which forms nearly 65% of the total expenditure in floricultural production. The major emphasis in the Dutch flower industry now seems to be on labour saving, specialization, quality and postharvest handling technology.

United States of America. Since the greenhouse technology in the United States of America was introduced by European immigrants, the development of the floricultural industry there is very recent as compared to the European flower industry [5]. The development of floriculture in the USA was largely dependent on the mode of transportation, production costs and the quality of the flower produced. In early years, the production of carnations, roses, chrysanthemums and foliage plants was concentrated in and around New York, Philadelphia, Boston and Chicago.

With the development of air transportation facilities during the 1950's the outdoor production of cut flowers in areas like Florida, California and to a lesser extent in Texas became a reality. The savings in production costs was more than the added cost of transportation. The need to produce flowers of acceptable quality under a controlled environment, at a minimum cost, helped the State of California to develop into the flower capital of the world. According to the 1981 census, California produced 64% pompom chrysanthemums, 61% standard chrysanthemums, 69.0% spray carnations and 50% of the total number of the Hybrid Tea roses produced in the USA. Roses, owing to their delicate nature, and susceptibility to harsh climatic conditions (wind, storm, frost) and inherently low vase life, continued to be produced in the north eastern states that are close to the major marketing centres.

The cultivation of carnations provides further evidence for the importance of production costs and quality in this crop in determining the total area of cultivation. In the 1950's, due to the higher cost of production of greenhouse-grown carnations and increasing competition from the Colombian carnation industry, the carnation production areas were shifted to Colorado, which has nearly perfect growing conditions for producing quality flowers. According to the 1980 census, Colorado produced 25% of the total carnations.

Roses and pot plants are still produced in close proximity to major markets. The competition from Holland, Israel, Kenya and Mexico is increasing in the rose industry. The production of foliage plants and other potted ornamentals in gaining in importance because of competition from neighbouring countries for the American fresh flower market.

1.2 Future Trends in the Flower Industry

The consumer demand for cut flowers and ornamental plants will continue to exceed production. The market for these products will expand at a much faster rate with more efficient marketing methods. The sale of floral products at the mass market level, discount and food stores will play a major part along with the traditional full service florist's shops. Organizations such as the Society of American Florists, Florafax, Florists' Transworld Delivery (FTD) etc. will continue to play a vital role in delivering quality goods and services to the consumer.

The major emphasis when buying flowers and other floral products will be on quality. Improved loss reduction biotechnologies, such as the development of maturity standards, the establishment of grades and standards, for different floral products, packaging technology, storage in controlled and modified atmosphere, low pressure and low temperature storage, use of bud opening and floral preservative solutions and improved marketing systems will become absolute necessities in the floral industry. Growers will face increasing pressures to produce better-quality products. The consumer desires concise information from the retailer on the care and handling of different products. Appropriate training at the retail level in care and handling of floral products will help this industry to flourish.

The growing communications system and the ease with which the improved technologies can be imported from developed countries will increase the competition from the countries in more tropical climates, where high quality cut flowers can be produced outdoors at a much lower cost. The flower grower will have to be more and more specialized in terms of both crops produced and the production of floral products for specific markets (mass market, retail market, full florist market). The quality of the product will determine the final success of the particular floral operation. Newer introductions in potted and bedding plants will receive greater attention in the future. In addition, smaller-sized pots (arrangement packs or 3–4" pots) and miniature flowers will be in greater demand, as compared to premium-sized pots or cut flowers with longer stem and larger flowers.

1.3 Quality in Flowers and Ornamentals

Quality in horticultural products is assessed by the relative values of several characteristics which, considered together, will determine the acceptability of the product to the buyer and the ultimate consumer [77]. There are several definitions for quality in floricultural crops, only the most important being quoted here.

Staby et al. [3] defined quality as the attributes of floral commondities that relate to their beauty and/or usefulness. Quality in horticultural products is an elusive term, upon which much depends but about which little is known. According to Arthey [77], acceptability is an important aspect and is included in many definitions of quality. Kramer and Twigg [78], for example, defined quality as the composite of those characteristics that differentiate individual units of the product and are significant in determining the degree of acceptability of that unit by the buyer.

Although most of these definitions imply contribution of several different attributes, general quality cannot be assessed accurately without separating these properties and measuring each of them individually.

Different products require different criteria for an assessment of their quality. Apart from quantitative (net weight, length of stalk, number of leaves, leaf and flower size) and qualitative (freedom from diseases, mechanical damage and condition of flower), hidden and sensory characteristics (appearance, colour and size) also determine the ultimate quality of floral products. For example, the criteria of cut carnations set forth by the Society of American Florists are as follows:

Flower:	size
	condition
	maturity
	shape
	longevity
Stem:	size
	shape
Absence of defects	

Although the relative significance of each of these characteristics determines the overall quality of the floral products, there is often a bias towards the more important quality component, longevity, in the case of cut flowers and potted foliage plants. Colour, texture, absence of defects, size, grading and fragrance constitute other important criteria of floral quality [77].

1.3.1 Components of Quality

Postharvest quality characterics, most importantly flower longevity, change during the handling, transportation, and marketing operations. The three major components of quality at the time of purchase are [77]: (1) appearance, (2) chemical, and (3) anatomical. Appearance includes size, shape (form), surface cleanliness, colour and condition (freshness). Optimum size of a product changes with time, depending upon the purchasing ability of the consumer. In recent years, there has been an increased demand for medium to small (miniature) types of flowers and potted plants. The mode of utilization of a commodity also influences its size, e.g. long-stemmed or large flowers and foliage plants may not be useful for flower arrangements or as house plants. The form or shape relates to consumers' preconceived notions regarding how the floral crops should look. Whereas in the USA, roses are preferred in their bud-harvested stages, and carnations and chrysanthemums are generally accepted fully opend, in Europe, bud-harvested floral crops are considered of high quality.

Physical blemishes due to insects, diseases, and pests, and/or certain physiological disorders present on the surface of flowers and ornamental plants markedly reduce their quality. High quality products must be clean and free of insects, pathogens, and residues of pesticides and other chemicals used to preserve their quality.

Colour has a major influence on consumer acceptability, more even than flavour does in most edible horticultural commodities, like fruits and vegetables. Significant amounts of sprays and tints are employed in the floricultural industry to protect flower colours and to fulfill specific consumer needs. Intangible quality characteristics, like freshness and absence of damage, refer to the subjective evaluation of term "condition" of the product and cover those factors or components of floral quality not included under the heading of size, form, surface, cleanliness and colour [3].

Very few floral crops were grown for their odour or fragrance and this chemical factor influencing flower quality had become less important in the recent past. Breeders, in fact, even tried to eliminate the fragrance factor in lieu of more desirable quality traits. However, fragrance is regaining importance in the floricultural industry.

Anatomical or physical characteristics of flowers such as crispness, toughness etc. are related to the texture of the product and have direct bearing on their keeping quality. While normally turgidity aids the freshness of cut flowers, sometimes slight wilting may actually improve their longevity and postharvest quality. Allowing standard chrysanthemums to wilt slightly during their last 2 to 3 weeks' period prior to harvest increases their keeping quality [3]. However, foliage and flowering potted plants should not be allowed to wilt, especially at the consumer level, because it may result in severe damage due to soluble salts accumulation.

Several preharvest, harvest and postharvest factors including genetic or inherent, climatic or environmental (light, temperature, relative humidity, pressure and composition of air) and managemental (soil condition, nutrition, fertilization, irrigation, plant protection etc.) factors, influence the postharvest quality and longevity of cut flowers and other ornamental plants [1, 78]. The effects of these factors are discussed in detail under individual crops.

1.4 Postharvest Losses and Problems of Handling in Flowers and Ornamentals

Owing to higher moisture content, ornamentals are highly perishable and, as such, are difficult to dry and preserve as dehydrated products. Furthermore, dried floral products are often not accepted by the consumers. Owing to their delicacy and tenderness, flowers and ornamentals are more susceptible to mechanical and physical damage and infection by diseases and pests during and after harvest. Even after being detached from the mother plant, the cut flowers are metabolically active and carry on all life processes at the expense of stored reserve food in the form of carbohydrates, proteins and fats, limiting their longevity to less than a few days. Certain flowers are inherently short-lived after harvest and pose unique problems in handling and marketing.

1.4.1 Extent of Postharvest Loss

According to a report of the National Floriculture Conference on *Commodity Handling* held at the Ohio State University in 1976 approximately 20% of all floral crops are unsaleable due to improper handling [4]. The economic value of loss of flowers in the USA increased from \$100 to \$134 million during the period from 1970 to 1975. This estimate did not include the quantity of crop that was never harvested, which could be over 10%, depending upon the produce and producer [4]. If the shrinkage losses that occur all along the marketing channel are added, the potential economic value of the total loss could be well over 50% of the farm value.

In a recent survey [12] on the credit (refunds) trends in a major California flower wholesaler, seasonal fluctuations noted in the volume of credits as a percentage of sales over a 4-year period accounted for nearly 1.4% of the total sales. The breakdown of credits by source indicated that by far the majority (80%) of the credits requests were related to poor quality of flowers on arrival. This was primarily attributed to the intrinsic poor quality of flowers (63.0%) and to a lesser extent to delays and temperature extremes (17%). These credit requests were in addition to the usual postharvest losses due to shrinkage. Careful handling of flowers could reduce such losses to a minimum.

Apart from the economic implications, shrinkage reduces the productive capacity of the industry and consumer purchasing. In almost all cases, the economic losses are ultimately passed on to the consumers, without providing them with a quality product [4]. This will affect the flower trade in the long run, and the utility creation process and the competitive position of cut flowers relative to other horticultural commodities.

1.4.2 Handling Problems Related to Flowers and Ornamentals

Present-day practices of harvesting, packaging, storage, transportation, marketing, advertising and design are examples of the application of modern scientific principles or biotechnologies in fluoriculture. There are, however, variations among the growers in employing the knowledge of improved handling techniques for the perishables. The growers, retailers and wholesalers need to be educated to ensure proper handling of flowers and ornamentals during and after harvest.

Potted plants pose an entirely different set of problems in their handling. Improper handling of these ornamentals after harvest reduces their quality and longevity, markedly affecting consumer acceptability of these products. Often times, the consumers do not realize that the product was damaged only during its post-production period, but visualize the product as having been produced with intrinsic defects.

1.4.3 Recent Developments in Loss Reduction Biotechnologies in Flowers and Ornamentals

Improved postharvest handling of floral crops and potted ornamental plants to maintain the high quality of the produce during transit has increased the international movement of these products in the recent years. Staby and Robertson [3] have summarized the following loss reduction technologies to reduce shrinkage losses in cut flowers and other foliage plants.

1. Selection of quality species and cultivars [3–4, 12–14].
2. Standardization of harvest standards [14–28].
3. Preshipping chemical treatments [28–38, 75].
4. Packaging techniques [3, 39–42].
5. Regulation of storage environment
   a. Temperature
      i) Precooling [43–48].
      ii) Mechanical refrigeration [49–50].
   b. Atmospheric composition and pressure
      i) Controlled atmosphere [60–61].
      ii) Modified atmosphere [55–59].
      iii) Hypobaric storage [51–54].
6. Use of floral preservatives and bud-opening solutions [62–65].
7. Mode of transportation [66–76].

References

1. Halevy AH, Mayak S (1979) Senescence and postharvest physiology of cut flowers, Part 1. In Janick J (ed) Hortic Rev, Vol 1, AVI Westport, Conn, pp 204–236
2. Zieslin N, Kohl HC Jr., Kofranek AM, Halevy AH (1978) Changes in the water status of cut roses and its relationship to bentneck phenomenon. J Am Soc Hortic Sci 103–176
3. Staby GL, Robertson JL (1982) International movement of cut flowers. HortScience 17:729

4. Staby GL, Robertson JL, Kiplinger DC, Conover CA (1976) *Proc Natl Flor Conf on commodity handling*, Ohio Flor Assoc (Columbus)
5. Janick J (1979) Horticulture's ancient roots. HortScience 14:299
6. Halevy AH, Kofranek AM (1977) Siver treatment of carnation flowers for reducing ethylene damage and extending longevity. J Am Soc Hortic Sci 102:76
7. Benschop M, De Hertogh AA (1969) Postharvest development of cut tulip flowers. Flor Rev 145 (3758):24, 62, 65
8. Mayak S, Halevy AH, Sagie S, Bar-Yoseph A, Bravdo B (1974) The water balance of cut rose flowers. Physiol Plant 32:15
9. Meeteren U Van (1978) Water relations and keeping quality of cut gerbera flowers. I. The cause of stem break, Sci Hortic 8:65
10. Wilberg B (1973) Physiologische Untersuchungen zum Knicker-problem als Voraussetzung für die Selektion haltbarer Gerbera-Schnittblumen. Z Pflanzenzucht 69:107
11. Nelson PV (1977) Greenhouse operation and management. Ruston, USA
12. Johnston JE, Reid MS (1981) Losses of flowers postharvest. A commercial point of view. Acta Hortic 113:69
13. Halevy AH, Mayak S (1975) Hormonal regulation of rose petal senescence, Proc 2nd Int Symp on Plant Growth Regulators, Publ House, Bulgarian Academy of Sciences, Sofia, pp 240–247
14. Cheng L (1965) Harvesting and handling of carnations at tight buds, MS Thesis, Colorado State Univ Ft Collings, 1965
15. Goszczynska D, Rudnicki RM (1978) The effect of silver nitrate on bud opening and senescence of ethephon-treated carnation buds. Acad Polonaise Sci Bull Ser Des Sci Billogigues 25:137
16. Goszczynska D, Nowak J (1979) The effect of growth regulators on quality and vase life of dry-stored carnations. Acta Hortic 91:143
17. Maxie EC, Farnahm DS, Mitchell FG, Parsons RA, Sommer NF (1973) Effectiveness of crushed ice in maintaining bud temperatures in containers of Forever Yours roses held 24 hours at room temperature. Roses Inc Bull pp 71–74
18. Anon (1969) Bud-cut carnations nip shipping costs. Agric Res (Wash DC) 17:12
19. Guilfoy RF, Lundquist AL (1970) Transport and handling of carnations cut in the bud stage-potential advantages. Proc Soc Am Florists pp 7
20. Hardenburg RE (1967) Bud cut carnations .... nip shipping costs. Agric Res (Wash DC) 17:12
21. Heide OM, Oydvin J (1969) Effects of 6-benzylamino purine on the keeping quality and respiration of glasshouse carnations. Hortic Res 9:26
22. Lancaster DM (1974) Effects of preharvest environment on keeping life of cut carnations. Bull Colorado Flower Grow Assoc 201:292
23. Mayak S, Dilley DR (1976) Effect of sucrose on response of cut carnation to kinetin, ethylene and abscisic acid. J Am Soc Hort Sci 101:583
24. Kofranek AM, Shepard P, Kubota J (1972) Seasonal bud opening of Albatross mums. Flor Rev 151:22, 59
25. Parups EV, Voisey PW (1976) Lignin content and resistance to bending of the pedicel in greenhouse-grown roses. J Hortic Sci 51:253
26. Kohl HC (1961) Rose neck droop. Bull Calif State Flor Assoc 10:4
27. Marousky FJ (1971) Handling and opening bud-cut chrysanthemum flowers with 8-hydroxyquinoline citrate and sucrose. US Dept Agric Mkting Res Rept 905
28. Hardenburg RE, Vaught HC, Brown GA (1970) Development and vase life of bud-cut Colorado and California carnations in preservative solutions following air shipment. J Am Soc Hort Sci 95:18
29. Kofranek AM, Kubota J (1973) Carnation production and quality as affected by cutting methods and spacing. Flower Nurs Rept Calif p 2
30. Halevy AH, Myayak S (1974) Improvement of cut flower quality, opening and longevity by pre-shipment treatments. Acta Hortic 43:335
31. Kofranek AM, Halevy AH (1972) Conditions for opening cut chrysanthemum flower buds. J Am Soc Hortic Sci 97:578

32. Veen H (1979) Effects of silver on ethylene synthesis and action in cut carnations. Planta 145:467
33. Veen H, Geijn S Van De (1978) Mobility and ionic form of silver as related to longevity of cut carnations. Planta 140:93
34. Veen H (1979) Effects of silver salts on ethylene production and respiration of cut carnations. Acta Hortic 91:99
35. Dimalla GG, Staden J Van (1980) The effect of silverthiosulphate preservative on the physiology of cut carnations. I. Influence on longevity and carbohydrate status. Z Pflanzenphysiol 99:9
36. Means K (1981) Silverthiosulphate! Supermarkets can tap into long carnation life. Flower Marketing 88(9):88
37. Reid MS, Farhoomand MB, Paul JL, Kofranek AM (1979) Extension of cut flower longevity by pulsing with silverthiosulphate. HortScience 14:445
38. Reid MS, Paul JL, Farhoomand MB, Kofranek AM, Staby GL (1980) Pulse treatments with the silver thiosulphate complex extend the vase life of cut carnations. J Am Soc Hortic Sci 105:25
39. Hinsch RT, Rit RE (1972) Cost and performance of full telescope corrugated boxes for shipping Colorado-grown carnations. US Dept Agric Res Ser Pub No 52
40. Rit RE, Hinsch RT (1972) New shipping containers for cut roses-costs and performance. US Dept Agric Res Ser Pub No 52:11
41. Marousky FJ (1971) Effects of temperature, container venting, and spike wrap during simulated shipping and use of floral preservative on subsequent floret opening and quality of gladiolus. Proc Trop Region Am Soc Hortic Sci 15:216
42. Molenaar WH, Nieuwenhuizen GA Van (1976) Geisoleerd-transport vooregekoelde snijbloemen. (Insulated transport of pre-cooled cut flowers). Vakbl Bloemist 31(41):74
43. Farnham D, Bryne T, Marousky F, Durkin D, Rij R, Thompson J, Kofranek A (1979) Comparison of conditioning, precooling, transit method and use of a floral preservative on cut flower quality. J Am Soc Hortic 104–483
44. Thompson JF (1980) Precooling cut flowers means better quality. Gladio Grams 638:11
45. Rij RE, Thompson JF, Farnham DS (1979) Handling precooling and temperature management of cut flower crops for truck transportation. US Dept Agric Adv Agric Tech 5:26
46. Holmes ES, Myers JM (1956–57) Precooling and transit refrigeration of gladioli. Annu Rept Fla Agric Expt Sta pp 38–40
47. Holmes ES, Myers JM, Sheeham TJ (1958–59) Gladiolus preceooling and transit investigations. Annu Rept Fla Agric Expt Sta pp 5–46
48. Sheeham TJ (1959–60) Gladioli precooling and transit investigation. Annu Rept Fla Agric Expt Sta pp 126
49. Wiersma O, Boer WC (1970) Vacuumkoelen Van snijbloemen te aalsmeer, Sprenger Inst Rapport, 1928
50. Wiersma O, Boer WC (1951) Vacuumkoelen en transport van snijbloemen. Sprenger Inst Rapport, 1808
51. Dilley DR, Carpenter WJ, Burg SP (1974) Principles and application of hypobaric storage of flowers. Bull Mich State Univ Agric Expt Sta 6644:1–18
52. Dilley DR, Carpenter WJ (1974) Principles and application of hypobaric storage of cut flowers. Acta Hortic 41:249
53. Patterson DS (1975) The effects of preservative solution in hypobaric storage on chrysanthemum flower longevity. MS Thesis, Wash State Univ, Pullman
54. Staby GL, Cunningham MS, Kelly JW, Konjoian PS, Holstead CL (1981) Low pressure storage of roses and carnations. HortScience 16:404
55. Thornton NC (1930) The use of carbon dioxide for prolonging the life of cut flowers with special reference to roses. Am J Bot 17:614
56. Longley LE (1934) Some effects of storage of flowers in various gases at low temperature on their keeping qualities. Proc Am Soc Hortic Sci 30:607
57. Neff MS (1942) Effects of storage conditions on cut roses. Bot Gaz 103:794

58. Thornton NC (1934) Carbon dioxide storage. VII. Changes in flower colour as evidence of the effectiveness of CO_2 in reducing the acidity of plant tissue. Contrib Boyce Thompson Inst 6:403
59. Tinga JH (1956) The effect of modified atmosphere storage at low temperature and treatments after low temperature storage which affect the keeping quality of cut flowers. Ph D Dissertation Cornell Univ
60. Hanan JI (1966) Controlled atmosphere storage of carnations. Bull Colorado Flower Grow Assoc 193:1
61. Asen S, Parsons CS, Stuart NW (1964) Experiments aimed at prolonging narcissus display life. Flor Rev 134 (3472):25, 69
62. Kofranek AM (1976) Opening flower buds after storage. Acta Hortic 64:231
63. Kofranek AM, Halevy AH (1976) Sucrose pulsing of gladiolus stems before storage to increase spike quality. HortScience 11:572
64. Kofranek AM, Halevy AH, Kubota J (1975) Bud opening of chrysanthemums after long term storage. HortScience 10:378
65. Robertson JL, Staby GL (1976) Economic feasibility of once over bud harvest of standard chrysanthemums. HortScience 11:159
66. Wiersma O, Boer WC (1973) Transport temperaturen on vacuumkoelen van snijbloemen (westland). Sprenger Inst Rapport 184:7
67. Besemer ST (1961) Shipping methods used for fresh cut flowers evaluated in transcontinental trials. Calif Agric 15:6
68. Harvey JM, Uota M, Segall RH, Ceponis MJ (1963) Transit times and temperatures of transcontinental cut flower shipments. US Dept Agric Mkting Res Mkting Res Rept 592
69. Maxie EC, Mitchell FG (1966) Air transport of perishables. HortScience 1:86
70. Halevy AH (1975) Checking methods of sea transportation of carnation, gladiolus and statice (in Hebrew). Annu Rept Hebrew Univ Ornamental Hortic pp 46–48
71. Mayak S, Gvili A, Halevy AH, Zafrer R (1969) Shipping cooled flowers by boat (in Hebrew). Annu Rept Hebrew Univ Ornamental Hortic pp 65–68
72. Mayak S, Halevy AH (1972) Checking the alternative methods of sending flowers by sea (in Hebrew). Annu Rept Hebrew Univ Ornamental Hortic pp 92–95
73. Vega EA (1964) Experiments on the transport of cut flowers by ship. Preliminary Rep Natl Univ Inst Agric Volcani Inst Dept Agron Garden Crops 474
74. Paulin A (1977) Le transport des fleurs coupes. Rev Gen Froid 1:55
75. Halevy AH, Mayak S (1974) Transport and conditioning of cut flowers Acta Hortic 43:291
76. Halevy AH, Byrne T, Kofranek A, Farnham D, Thompson J, Hardenburg R (1978) Evaluation of postharvest handling methods for transcontinental truck shipments of cut carnations chrysanthemums and roses. J Am Soc Hortic Sci 103:151
77. Arthey VD (1975) Quality of Horticultural Products. Butterworth and Co Ltd, London (1970) pp 1–23
78. Kramer A, Twigg BA (1970) Quality control for the Food Industry. Vol. I. AVI Westport, Conn

Chapter 2 Senescence of Flowers and Ornamentals – Basic Principles and Considerations

2.1 Definition and Terminology

Senescence is one of the least well-defined steps in biological development. The gross change of a series of changes leading finally to death of an organism has been referred to as senescence, Sacher [1] defined senescence as the final phase in the ontogeny of the organ in which a series of normally irreversible events is initiated that leads to cellular breakdown and death of the organ. Senescence of higher plants is classified into three major types [2, 4]: (a) population senescence (e.g. annual plants), (b) organism or individual plant senescence, and (c) determinate organ senescence (e.g. leaves, fruits, flowers, petals etc.).

Determinate organs such as flowers most probably undergo a genetically programmed senescence, which is independent of specific "ageing genes" and is devoid of any special genetical mechanism of senescence. Sacher [1] described determinate organ senescence on the basis of general gene action of normal plant development. Experimental evidence is lacking to demonstrate the presence of specific "ageing genes". Ample evidence, however, indicates that the increased activity of several enzymes that function during plant development plays an important role during organ senescence [1].

Leopold [2, 3] defined senescence as the deteriorative processes that are the natural causes of death; ageing, by contrast, refers to processes accruing maturity with the passage of time. Ageing thus includes a much wider span of physiological processes, which may either weaken the organism or be neutral with respect to the capability of the biological organism or be neutral with respect to the capability of the biological organism to survive. Senescence, in contrast, refers to the changes providing for the endogenous regulation of death [4].

"Longevity", a correlate of ageing and senescence, is linked to flowering and fruiting, as is evident in the case of annual plants. The enormous ranges of longevity over the plant kingdom (ranging from a few weeks to several thousand years) has been related to the dynamics of organism turnover. The rate of evolutionary change is represented as the product of the extent of variability of a population of organisms times the turnover rate; those having a high degree of variability between individuals and also a short life span have the greatest capacity for evolutionary change. In general, the shortest lived plant species tend to have longevities limited by programmed senescence. The perennials, which have much less precise limits of longevity, die of gradual process of ageing [4].

2.2 Structural, Biochemical and Metabolic Changes Associated with Senescence of Flowers and Ornamentals

2.2.1 Ultrastructural Changes

Mayak and Halevy [5] and Halevy and Mayak [6] have reviewed and updated the work done on ultrastructural, biochemical and biophysical changes during petal senescence. Extensive studies carried out on ultrastructural properties of petal senescence in the ephemeral corolla of the morning glory indicated invagination of the tonoplast as the first observed sign of senescence. This in turn suggested involvement of autophagic activity of the vacuole, representing the lysosomal cell compartment [9]. According to Matile and Winkenbach [9], abolition of the compartment of the vacuole and the consequent release of hydrolases resulted in the death of the cell. The presence of cytoplasmic material such as disintegrated mitochondria and different types of membranes in the ageing vacuole support this hypothesis. The breakdown of the tonoplast is later followed by autolysis of the cell.

During maturation and senescence first free single ribosomes disappear followed by those in aggregate clusters and those attached to the endoplasmic reticulum [9]. Vascular occlusions decrease uptake of water and consequently reduce vase life of cut flowers [7]. Burdett [7] reported that stem sterilization or 8-hydroxyquinoline solutions decreased solution flow resistance by controlling microbial growth. Microscopic examination of longitudinal tangential sections of rose stems have revealed vascular occlusions due to microbial growth and gum deposition [8]. This indicates a loss of membrane integrity, causing increased permeability and leakage [10].

An increase in apparent free space and membrane permeability during the senescence of several flowers has been demonstrated [11–16]. Borochov et al. [17, 18] observed a sharp increase in microviscosity of the plasmalemma during ageing of intact rose flowers, cut flowers or isolated petals. It was further observed that the increase in microviscosity corresponded to an increase in the ratio of free sterol to phospholipid. The free sterol content of the flower remained unaltered during senecence, but the content of phospholipid was reduced. Halevy [10] attributed this to reduced synthesis and increased hydrolysis by phospholipase A, which decreased the level of phospholipids. Beutelmann and Kende [19] noted that such a decline in phospholipids was observed in the *Ipomoea corolla*, even before visible signs of ageing were perceptible. Simons [97] postulated that the decrease in the level of phospholipids enhanced the permeability of the plasma membrane, causing cell leakage. Beutelmann and Kende [19] found that during senescence, there was a 40% decrease in the level of fatty acids esterified to phospholipids, which corresponded to the loss of phospholipids during the same period; no significant changes were found in the ratio of the fatty acid components. The level and composition of free fatty acids remained unchanged during senescence.

The symptoms of loss of fresh weight of the flower tissue, such as drying and shrivelling, are conspicuous in the final phase of senescence, the loss of water occurring even when the ageing petals of cut flowers are held in water, indicating a loss of membrane integrity and consequent increased permeability and leakage [10].

According to Thimann [20], the first microscopically visible changes in leaf senescence are those in the chloroplasts. The chloroplasts lose their characteristic starch deposits, due to reduction of the bulk of starch into sugars. Mitochondria first become distorted, then either disappear or at least shrink. Cells of the same leaf may senesce at different rates: in some the whole plastid envelope disappears, while in others this membrane is intact even after 12 days. Eventually, however, the tonoplast and plasmalemma both disappear. The hydrolases present in the vacuole attack the cytoplasm. The dissolution of weakening membrane produces great changes in the permeability of senescing cells, leading to leakage of solutes [1]. Delayed or reduced leaf senescence may be associated with reduced levels of the leaf peptide hydrolases or with a postponement of their rate of appearance [20].

2.2.2 Biochemical Changes

Respiration and enzymatic hydrolysis of cellular components are the two major biochemical and metabolic events occurring in the senescing flowers and petals of tulip [21] and orchid [22, 42–44], carnation [45, 46], rose [45] and chrysanthemum [45]. The enhanced peroxidase activity was associated with an increse in the level of peroxides and free radicals, which reacted with cellular constituents [23], and are probably involved in promotion of senescence [24, 25)and in the production of ethylene [26]. A delay in the senescence of carnation flowers was demonstrated by treating them with free radical scavengers such as sodium benzoate [27]. Significant increases in the activities of ribonuclease, deoxyribonuclease, and hydrolases of cell wall polysaccharides of *Ipomoea* [28] and of ribonuclease in roses [29, 30] were reported.

During the course of petal senescence, a decrease in the level of the following macromolecular components was noticed: starch [31], cell-wall polysaccharides [28], proteins [32–35], and nucleic acids [9]. The main constituents of the sugar pool of mature petals of carnation and rose were the reducing sugars, rather than the non-reducing sucrose [15, 36], the majority of the transformation in sugars being accompanied by hydrolysis of starch [31].

Insoluble pectins, present as the main constituent of the cell wall, were found in different amounts in two varieties of carnation having different vase life. Astor, a red short-lived cultivar, was characterized by 25% lower pectin content than the yellow long-lasting cultivar, Alice [41].

The marked development of senescence in flowers was associated with a variety of changes, such as blueing of red petals, decrease in protein content, and increase in ribonuclease activity [29, 85, 92, 95].

Since senescence of cut flowers is closely related to depletion of energy required for synthetic reactions, an exogenous supply of sugar has been recommended as the most efficient means of delaying the onset of senescence [49, 56]. Glucose supply prevents a sharp decrease in the amount of soluble proteins in petals [35], whereas Kaltaler and Steponkus [56] postulated that the principal effect of applied sugars is to maintain the structure and function of mitochondria. The rapid decline in the rate of water uptake and the relatively sustained rate of transpiration when cut flowers are placed in water also leads to loss of turgor. This has been attributed to vascular blockage at the stem base reducing water uptake by the stem [8].

2.2.3 Metabolic Changes

Internal metabolic changes in respiration of cut flowers have been reported by several workers [49–58]. The rate of respiration in many cut flowers reaches its peak at the time of opening of flowers, and decreases as the flowers mature and senesce. Later, there is a second dramatic increase in respiration over a relatively short period, followed by a final decline [5]. According to Mayak and Halevy [5], the second peak in the respiration drift signifies the last phase of senescence. It has been considered to be an analogue to the climateric rise in respiration of many fruits. The cyanide insensitivity of respiratory rise in certain flowers (e.g. Araceae), indicates the formation of free radicals with high oxidation potential, which promote senescence in many plant tissues and are associated with increased sensitivity of the tissue to ethylene [23, 24]. The second peak in respiration of flowers may be employed to assess the effectiveness of senescence-retarding substances, provided this peak in respiration reflects internal metabolic changes associated with ageing. Chemicals that delay the occurrence of the second respiratory peak have been reported to increase vase life of flowers [49, 50]. Hew et al. [51] demonstrated a rhythmic pattern of respiration in orchid flowers, CO_2 production starting as soon as the flowers opened. The period between amplitudes was about 24 h, and was not influenced by continued darkness. The amplitude of the rhythm was slightly dampened in cut flowers, but it was partially enhanced by supplying sucrose externally.

Weinstein [36] proposed that the onset of hydrolysis of cellular components such as proteins and carbohydrates is initiated in response to the depletion of free sugars used up in respiration, to supply alternative respirable substrates (e.g. carbon skeletons of amino acids or keto acids). This hypothesis is supported by the observation that an exogenous supply of sugars delays the onset of excessive degradation of protein [48].

Evidence forwarded by Kaltaler and Steponkus [56] indicated that the gradual decline in respiration and decrease in respiration efficieny of rose petals resulted from the progressive inability of the mitochondria to utilize the substrate. A change in respiration control ratio of related mitochondria probably affected their integrity, thereby reducing the functional capacity. The respiration of isolated mitochondria from cut flowers could be controlled over longer periods of time. The main effect of applied sugar in prolonging vase life of roses was to protect the structure and function of mitochondria. However, the influence of sugar on these organelles may not be the specific effect, but the general protective effect on membrane integrity [59].

2.2.4 *Changes in Pigments*

Discolouration or fading of colour is a common sympton of many senescing flowers. The carotenoids and anthocyanins, two major classes of pigments responsible for different colour in flowers, change significantly during the development and maturation of plant organs. Simpson et al. [60] monitored changes in the composition of carotenoid in *Strelitzia raginae* flowers throughout the development of the plastids from small, colourless plastids through the green chloroplasts to large spindle-shaped chromoplasts. These authors noted an increase in the concentration of oxygenated carotenoids with age, similar findings being reported in roses [61].

Differential changes in the anthocyanin content of senescing flowers have been noted. While its level stays stable in certain flowers [62–63], in others it decline significantly [64], whereas in orchids anthocyanines are synthesized continuously.

Stewart et al. [65], noted that changes in the colour of senescing petals is significantly influenced by a change in the pH of the vacuole. Co-pigmentation with other flavonoids and related compounds is the decisive factor in determining the intensity of the colour in most flowers; in only a few cases is the colour caused by a very low (less than 3) or a very high (more than 7) pH affecting the anthocyanin per se. Asen et al. [66], found that the degreee of copigmentation is greatly influenced by even slight changes in pH. The blueing of red flowers with ageing and a concomitant increase in pH of senescing flowers have been reported in roses [32, 67], lathyrus [62], morning glory [38] larkspur [66], geranium and other flowers [65], which was ascribed to the breakdown of proteins and the consequent increase in free ammonia [37, 40].

2.3 *Hormonal Regulation of Senescence of Flowers and Ornamentals*

Even though the members of each of the five groups of plant hormones viz., auxins, gibberellins, cytokinins, abscisic acid and other growth retardants and ethylene have been implicated in the regulation of flower senescence, the effects of ethylene have been investigated most thoroughly and extensively.

2.3.1 *Effects of Ethylene*

There are several reports of damage to flowers induced by ethylene [39, 68–71] such as premature senescence, wilting of the corolla etc. Mayak and Halevy [5] cited striking examples of effects of ethylene in flowers: (a) the in-rolling of carnation petals, often termed sleepiness [15], (b) fading and in-rolling of the corolla in *Ipomoea* flowers [72], (c) fading and wilting of sepal tips in orchids [73], and (d) induction of anthocyanin formation in both gynostemia and labella [74]. These well-defined symptoms are often utilized to assess the effects of ethylene quantitatively. Uota [70] observed that CO_2 levels of 7% to 20% prevented sleepines in carnation blooms that were exposed to an ethylene

Table 2.1. Competitive inhibition of ethylene-induced sleepiness by CO_2 [70]

Ethylene (ppb)	Sleepiness (%) at CO_2 concentration (%)				
	0	4	7	10	20
0	5.5[a]	1.3[a]	0.0[a]	0.0[a]	0.0[a]
125	62.7	6.8[a]	0.0[a]	0.0[a]	0.0[a]
250	89.0	12.3	1.3[a]	0.0[a]	0.0[a]

[a] Means followed by different letters are significant at 1% level.

concentration of 125 to 250 ppb for 24 h (Table 2.1). Even 4% CO_2 had an effect similar to the higher concentration. Higher levels of CO_2 (more than 2%) were required to reduce sleepiness of the blooms when ethylene was increased to 500–1000 ppb. The treatment of both the flower [15, 72, 75] and fruit tissues [1] with the externally supplied ethylene has been known to catalyze the synthesis of ethylene endogenously by the plant tissue which is generally referred as autocatalysis.

The treatment of young and mature flowers of carnation with ethylene resulted in a dual response viz., the in-rolling of petals, and enhanced ethylene synthesis [3]. Kende and Hanson [50] also reported that exposure of mature *Ipomoea* flowers (on day of opening) resulted in a similar dual response. The younger buds (a day before opening), however, responded only with the in-rolling of the corolla. The results of the experiments carried out by Kende and Hanson [50] indicated that an ethylene-generating system develops as an integral part of the ageing process in flower tissue. Although ethylene hastened ageing of morning glory flowers, it did not appear to play an obligatory role in flower senescence. A working hypothesis forwarded by these researchers stated that the ageing process in morning glory flowers commences before ethylene production begins, and according to this view, ethylene does not trigger ageing or senescence, but accelerates it. The development of in-rolling of carnation petal tissue associated with the endogenous production of ethylene can be prevented totally with 4% CO_2, which acts as a competitive inhibitor of ethylene [15, 76, 77]. The treatment with 4% CO_2 delayed but did not prevent in-rolling of the corolla in *Ipomoea* flowers [50]. Wide variations in ethylene response of different flowers are thus encountered. In *Ipomoea*, the in-rolling of the corolla does not appear to be well correlated with autocatalysis of ethylene, even though the latter accelerates the in-rolling process. In contrast, ethylene seems to trigger the in-rolling process in carnations [5]. Nichols [15] noted that 5% CO_2 was sufficient to counteract the toxic dose of 0.2 ppm ethylene for 2 days in carnation (Table 2.2). Baker et al. [27] reported that ethylene production by carnation flowers was inhibited slightly by ethoxy analogue of rhizobitoxine (Ro) and sodium benzoate on day 2 but the inhibition was not strong until day 5 for the Ro-treated samples and day 4 for the benzoate-treated sample.

Table 2.2. Survival of flowers after treatment with 0.2 ppm ethylene and 2% carbon dioxide for 2 days (mean of ten flowers per treatment) [15]

Treatment	Survival (days)		
	Expt. 1	Expt. 2	Expt. 3
Control	7.5	5.4	5.7
C_2H_4	1.7	0	0
$C_2H_4+CO_2$	6.5	6.1	6.3
L.S.D. at P = 0.05	1.2	0.4	0.5

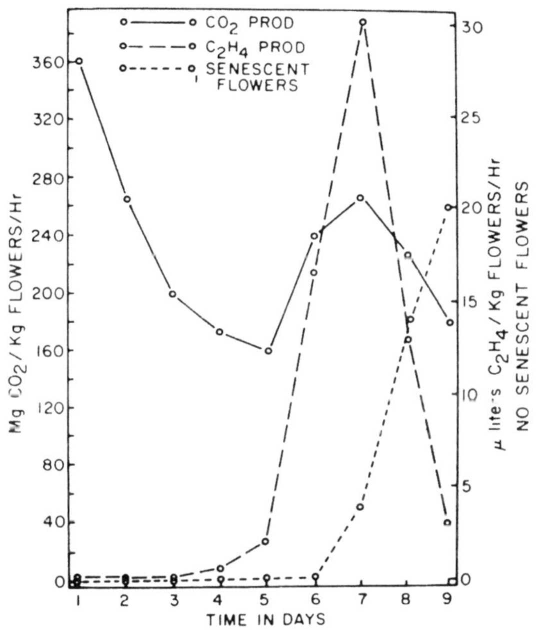

Fig. 2.1. Rates of respiration, ethylene production and senescence of flower of Improved White Sim carnation at 20 °C [78]

Maxie et al. [78] reported that the time course of ethylene production followed a typical profile composed of three distinct phases: (a) a low steady rate of ethylene production, (b) an accelerated rise of ethylene production to a peak, and (c) a decline in the ethylene production (Fig. 2.1). Different events associated with ageing of flowers and other ornamentals can be described with reference to these three phases [5]. The visual symptoms generally occur only in the second phase and are pronounced towards the end of this phase. The onset of the second phase thus signifies the terminal step of senescence. The transition of ethylene production to the second phase is well correlated with longevity of roses, as has been demonstrated in short-lived cultivars compared with the long-lived ones [79] (Fig. 2.2).

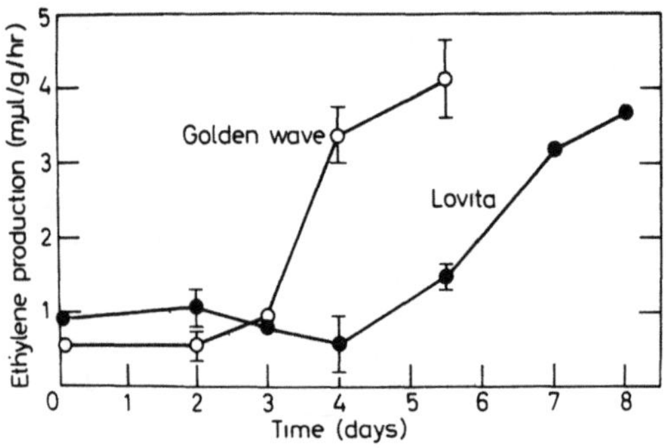

Fig. 2.2. Ethylene production by rose flowers. Long-lived cv. Lovita and short-lived cv. Golden Wave flowers were tagged at the young stage, and sampled on each of the following days [79]

Mayak and Halevy [5] stated that application of CO_2 to carnation during the first phase of ethylene production delayed the terminal phase of senescence, whereas the application of CO_2 at the start of the second phase was sometimes ineffective in delaying the transition into the auto-stimulation phase of ethylene production. The onset of the second phase can be delayed by CO_2 [76–77], low O_2 concentration [15] and hypobaric storage [76–77].

It is important to note that the application of ethylene does not induce ethylene synthesis in the young and early maturing phases of flower petals, although ethylene is constantly emanating in low amounts. Yet transition into the second phase does not occur. However, when the young flower tissues of carnation [80] are physically injured, ethylene production rises immediately. Later, after relatively short periods of time, it declines without developing typical systems of senescence.

The senescence of carnation has been known to be accelerated when the mature flower tissue is exposed to ethylene. The typical symptoms of senescence are evident even when the ethylene is ventilated away; the irreversible symptoms [81] are possibly associated with a rise in ethylene synthesis. The tissues of different maturity vary considerably in their response to the presence of ethylene; whereas young flower are not responsive, the mature ones respond and transition into the second phase is evident. The style has been shown to produce more ethylene than other parts. The auxin from pollen is transferred to the stigma where it spreads and triggers ethylene synthesis [75, 84].

According to Mayak and Halevy [5], and Halevy and Mayak [6] the entire process of ethylene-induced senescence is under programmed regulation or genetic control. The responding system, however, may be influenced by environmental factors such as temperature or water stress to cut flowers imposed by excessive drying and low temperature storage [82, 83]. All these conditions increase the sensitivity of the flower tissue to ethylene.

Experiments in which the osmotic concentration was maintained artificially also resulted in similar findings. The treatment of cut flowers with sugars and some mineral ions induced osmotic adjustment and extended the longevity of carnation and chrysanthemum by decreasing their sensitivity to ethylene.

2.3.2 Involvement of Other Plant Hormones

Plant hormones other than ethylene are also involved in the regulation of senescence of cut flowers and ornamentals. Mayak and Halevy [85] observed that the endogenous level of cytokinins in the petals of rose decreased as the flowers aged, and the level was lower in a short-lived cultivar than in a long-lived one. An external application of cytokinins delays senescence in several types of flowers [76, 85, 86]. These observations support the hypothesis that the reduction of the endogenous level of cytokinin is associated with plant senescence. Mayak and Halevy [30] also reported that application of kinetin retarded the loss of dry weight of the maturing flowers. The kinetin enhanced uptake of water probably by maintaining cell integrity.

The gibberellic acid was found to increase flower size and extended the longevity of carnations [87–88]. According to Gilbart and Sink [89], auxins play a central role in regulating senescence in poinsettia. The level of auxin decreased in two cultivars of poinsettia, the faster decrease taking place in the short-lived cultivar. The observation that the activity of IAA-oxidase and the level of hydrogen peroxide increased with ageing led Gilbart and Sink to propose that auxin probably delays senescence in carnation by inducing the synthesis of peroxidase, which then prevents the accumulation of free peroxide associated with ageing. According to Wulster et al. [90], IAA induces senescence in isolated carnation petals by increasing the duration and amount of ethylene production. The ability of petals to respond to IAA appeared to be a function of physiological age. These authors also noted that the IAA enhanced ethylene evolution only in specific portions of the petal.

Abscisic acid (ABA) has been assigned a significant role in the regulation of senescence in flowers; the endogenous level of ABA rises markedly as the petals senesce [5, 92]. Mayak and Dilley [77] reported that ABA shortened longevity of carnation flowers, and hypobaric ventilation, which enhances diffusion of ethylene and other gases from the tissues, largely negated this effect of ABA (Table 2.3). Gassing ABA-treated blooms with 3.8% CO_2 in air completely negated the effect of ABA, while 1%–2% CO_2 largely overcame the ABA influence. The inability of hypobaric ventilation to negate the effect of ABA was thought to be due to the fact that ABA stimulated the onset of

Table 2.3. Effects of abscisic acid (ABA) on carnation senescence in response to hypobraric ventilation with or without supplemental ethylene [77]

Treatment	Flower longevity (days)
Control	5.0
ABA 50 (micromolar)	3.5
Control plus hypobaric ventilation	9.3
ABA plus hypobaric ventilation plus ethylene	6.8
Control plus hypobaric ventilation plus ethylene	1.3
ABA plus hypobaric ventilation plus ethylene	1.6

0.2 atm. of 99.85% O_2 + 0.15% $CO_2 \pm 2.36$ µl/l ethylene autocatalytic ethylene production. ABA apparently sets the stage for the onset of accelerated ethylene production, since its effect is entirely masked by applying ethylene exogenously under hybobaric condition. Mayak and Halevy [91] noted that the level of ABA rose higher in a short-lived variety than in a long-lived one. In both carnations and roses, the external application of ABA accelerated senescence [77]. Halevy et al. [29] reported that by causing stomatal closure on the leaves, ABA reduced water loss from the flowering shoot, thereby delaying flower senescence.

It has generally been accepted that many plant processes, including senescence, are controlled through a balance between plant hormones interacting with each other and with other internal factors [5]. Ballantyne [93] reported that combined application of benzyladenine (BA) and auxin to daffodils was more effective in delaying senescence and respiratory climateric than the additive effect of the two hormones.

A hypothesis by Mayak and Halevy [91] states that ABA accelerates developmental processes associated with maturation and increases sensitivity of the tissue to ethylene. The tissue may respond to the existing concentration of ethylene without increasing its sensitivity to bring about an effect. The rise in the level of ABA, in response to treatment with ethylene observed in cut rose flow-

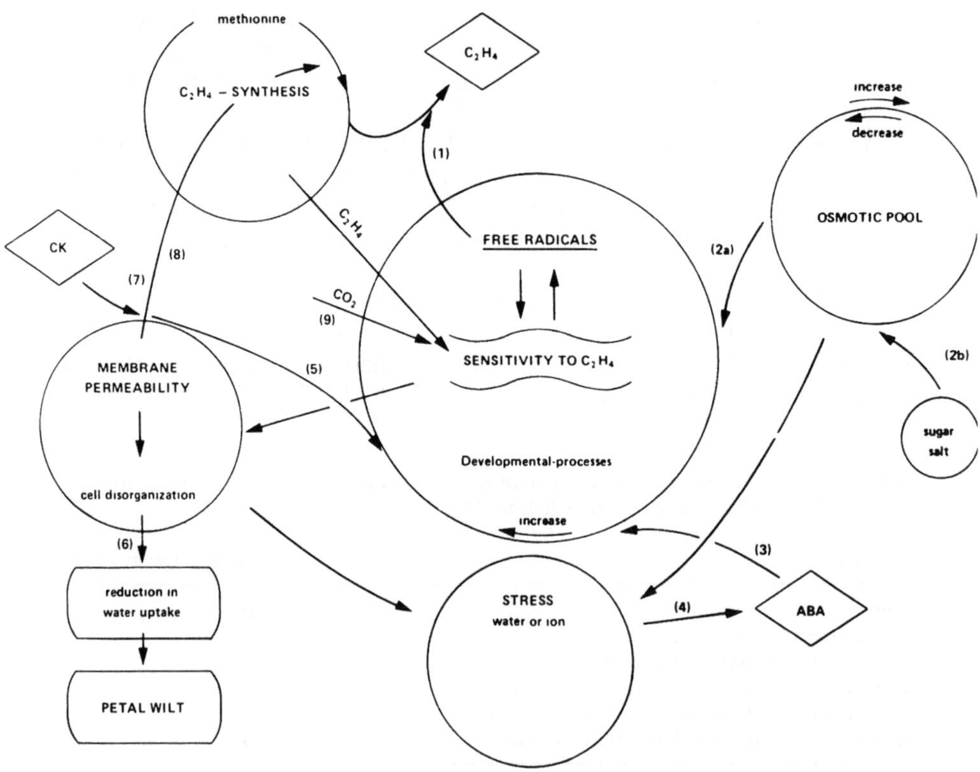

Fig. 2.3. A hypothetical scheme describing relations of plant hormones to flower senescence [5]

ers, was interpreted as a secondary result via an effect on membrane permeability and water stress imposed by ethylene.

Some of the effects of ethylene on the level of ABA appear to be mediated through modification of the membrane. The observation that kinetin delays wilting of cut roses by protecting cell integrity [30], supports the possibility of the reciprocal effects of cytokinin and ABA on each other's activity. The development of a stress situation could be delayed, and the level of ABA could be kept low by maintaining cellular integrity [5].

Based on a literature survey, Mayak and Halevy [5] proposed a hypothetical scheme summarizing some tentative generalizations concerning the processes and interactions with phytohormones (Fig. 2.3).

2.4 Shedding of Flowers and Ornamentals (Foliage)

Abscission or shedding of flowers and other ornamentals is the final phase of senescence, which may involve shedding of flowers, parts of flowers or the whole inflorescence. Much research work has been carried out on the abscission of leaves and fruits [94–95]. How the physiology and biochemistry of petal or flower abscission compares with this work is still under speculation.

The formation of an abscission layer, as with leaves, takes place during flower or bud shedding [96–97]. According to Esau [98], cell division usually does not precede petal shedding, no clear abscission layer is apparent, and petal shedding is caused by softening of the middle lamella. The latter has been reported to involve a rising activity of cell wall hydrolases, as in leaves [99].

External factors such as shaking, injury, adverse temperatures, certain plant gases and growth substances have been known to induce rapid petal shedding in sensitive species [100]. Ethylene promotes shedding of flower buds and petals in many flowers, including roses [68]. Carbon dioxide, which is generally antagonistic in its effects to ethylene, also promoted petal shedding. Pollination, especially fertilization, promotes petal abscission in several flowers, and can be attributed to a rise in ethylene production [73, 75].

There are several reports of promotion of abscission of flowers, buds and petals by ABA as is the case with leaf abscission [96, 99]. The level of endogenous ABA has been correlated with shedding of flowers of *Lupinus luteus* [100]. Whereas auxins retarded abscission of flower buds [97] and petals [99], the endogenous level of IAA coincided with seasonal fluctuation in flower bud shedding in begonia. Cytokinin was reported to be a better inhibitor of flower abscission than auxin, in roses [101].

Peterson et al. [102] examined relationships among water deficit, ABA content of leaf tissue and leaf abscission in intact *Ficus benjamina* L. (weeping fig). A strong inverse linear correlation existed between ABA content of leaves and plant water potential. However, leaf abscission was not found to be related with ABA content of leaves. The ABA content of leaves having a water potential of −0.5 bar was about 75-fold greater than that of leaves with a water potential of −8 bar. These results indicate that ABA does not independently regulate leaf abscission in *Ficus benjamina.*

2.5 Concluding Remarks

Senescence is a positive force in the ability of plant species to adapt to limitations of the environment, both climatic and competitive, and it can be a strong factor in the evolutionary adaptability of plants. While the mechanism of bud and flower senescence is still obscure, descriptive experiments conducted indicate that signals from the flowers and fruits are sent out to trigger the onset of death of plant organs or the overall plant. Mobilizing forces are probably involved in the senescence signals, some of which can be mimicked through application of plant hormones, such as kinetin [2].

References

1. Sacher JA (1973) Senescence and postharvest physiology. Ann Rev Plant Physiol 24:197
2. Leopold AC (1961) Senescence in plant development. Science 134:1727
3. Leopold AC (1975) Aging, Senescence, and turn over in plants. Bioscience 25:659
4. Leopold AC (1980) Aging and senescence in plant development. In: Thimann KV (ed) Senescence in Plants, CRC, Boca Raton pp 1–12
5. Mayak S, Halevy AH (1980) Flower senescence. In: Thimann KV (ed) Senescence in Plants, CRC, Boca Raton p 132
6. Halevy AH, Mayak S (1979) Senescence and postharvest physiology of cut flowers. part 1. In: Janick, J (ed) Hortic Rev, vol 1. AVI, Westport, Conn p 204
7. Burdett AN (1970) The cause of bent neck in cut roses. J Am Soc Hortic Sci 95:427
8. Lineberger RD, Steponkus PL (1976) Identification and localization of vascular occlusions in cut roses. J Am Soc Hortic Sci 101:246
9. Matile P, Winkenbach F (1971) Function of lysosomes and lysosomal enzymes in the senescing corolla of the morning glory (*Ipomoea purpurea*). J Exp Bot 22:759
10. Halevy AH (1981) Petal senescence at the cellular level. In: Second Int Symp on Postharvest Physiology of Cut Flowers Acta Hortic 113:151
11. Eliam Y (1965) Permeability changes in senescing tissue. J Exp Bot 16:614
12. Parups EV, Chan AP (1973) Extension of vase-life of cut flowers by use of isoascorbate-containing preservative solution. J Am Soc Hortic Sci 98:22
13. Hanson AD, Kende H (1975) Ethylene-enhanced ion and sucrose efflux in morning glory flower tissue. Plant Physiol 55:663
14. Mayak S, Vaadia Y, Dilley DR (1977) Regulation of senescence in carnation (*Dianthus caryophyllus*) by ethylene: Mode of action. Plant Physiol 59:591
15. Nicols R (1968) The response of carnations (*Dianthus caryophyllus*) to ethylene. J Hortic Sci 43:335
16. Sacalis JN (1975) Vascular blockage and its inhibition in cut rose flowers. Acta Hortic 41:159
17. Borochov A, Halevy AH, Shinitzky M (1976) Increase in microviscosity, with aging in protoplast plasmalemma of rose petals. Nature 263:158
18. Borochov A, Halevy AH, Borochov H, Shinitzky M (1978) Microviscosity of rose petal plasmalemma as affected by age and environmental factors. Plant Physiol 61:812
19. Beutelmann P, Kende H (1977) Membrane lipids in senescing flower tissue of *Ipomoea tricolor*. Plant Physiol 59:888
20. Thimann KV (1980) The senescence of leaves. In: Thimann KV (ed) Senescence in Plants, CRC Boca Raton, Florida, USA, p 86
21. Carfantan N, Daussant J (1975) Peliminary study of tulip protein during senescence. Acta Hortic 41:31
22. Trippi VS, Tran Thanh Van M (1971) Changes in the pattern of some isoenzymes of the corolla after pollination in *Phaleanopsis amabilis* Blume. Plant Physiol 48:506
23. Fridovich I (1975) Superoxide dismutase. Annu Rev Biochem 44:147

24. Brennan T, Frenkel C (1977) Involvement of hydrogen peroxide in the regulation of senescence in pear. Plant Physiol 59:411
25. Mishra SD, Gaur BK, Bedekar VM, Singh BB (1976) Isolation, identification and significance of free radicals in senescing leaves. Acta Bot Indica 4:131
26. Beauchamp C, Fridovitch I (1970) A mechanism for the production of ethylene from methional. The generation of the hydroxyl radical by xanthine oxidase. J Biol Chem 245:2641
27. Baker JE, Wang CY, Lieberman M, Hardenburg R (1977) Delay of senescence in carnations by a rhizobitoxine analog and sodium benzoate. HortScience 12:38
28. Wiemken-Gehring V, Wiemken A, Matile F (1974) Mobilization von Zellwandstoffen in den welkenden Blüten von *Impomoea tricolor* (Cav.). Planta 115:297
29. Halevy AH, Mayak S, Tirosh T, Spiegelstein H, Kofranek AM (1974) Opposing effects of abscisic acid on senescence of rose flowers. Plant Cell Physiol 15:813
30. Mayak S, Halevy AH (1974) The action of kinetin in improving the water balance and delaying senescence processes of cut rose flowers. Physiol Plant 32:330
31. Ho LC, Nichols R (1977) Translocation of ^{14}C-sucrose in relation to changes in carbohydrate content in rose corollas cut at different stages of development. Ann Bot (Lond) 41:227
32. Borochov A, Mayak S, Halevy AH (1976) Combined effects of acid and sucrose on growth and senescence of rose flowers. Physiol Plant 36:221
33. Parups EV (1971) Disc electrophoresis of proteins of senescing and fresh leaves and petals of certain ornamental plants. J Am Soc Hortic Sci 55:775
34. Paulin A (1971) Influence de la composition de la solution nutritive sur la teneur en divers acides amines libres et an ammoniac des petales de fleurs coupees. Ann Technol Agric 20:283
35. Paulin A (1977) Metabolism glucidique et proteique de la fleur d-oecillet alimente ou non avec une solution de saccharose. Acta Hortic 71:241
36. Weinstein LH (1951) Senescence of roses. I. Chemical changes associated with senescence of cut 'Better Times' roses. Contrib Boyce Thompson Inst 19:33
37. Dilley DR, Carpenter WJ (1975) The role of chemical adjuvants and ethylene synthesis on cut flower longevity. Acta Hortic 41:117
38. Asen S, Stewart RN, Norris KH (1977) Anthocyanin and pH involved in the color of 'Heavenly Blue' morning glory. Phytochemistry 16:1118
39. Beyer EM Jr (1977) $^{14}C_2H_4$: its incorporation and oxidation to $^{14}CO_2$ by cut carnations. Plant Physiol 60:203
40. Yazaki Y (1976) Co-pigmentation and the color change with age in petals of *Fuchsia hybrida*. Bot Mag 89–45
41. Jona R, Accati E, Mayak S (1981) Senescence processes as reflected in change in polysaccharidic cell wall components. In: Second Int Symp on Postharvest Physiology of Cut Flowers. Davis, California 21–25, July 1980. Acta Hortic 113:153
42. Tan TN, Hew CS (1973) Polyphenol oxidase activity in orchid flowers. J Singapore Nat Acad Sci 3:292
43. Lim SL, Chin TY, Hew CS (1975) Biochemical changes accompanying the senescence of *Arundina* flower. In: Biology in Society Singapore Academy of Science, p 18
44. Hsiang THT (1951) Physiological and biochemical changes accompanying pollination in orchid flowers. II. Respiration, catalase activity and chemical constituents. Plant Physiol 26:708
45. Parups EV (1976) Acid and alkaline inorganic pyrophosphatases in senescing flowers of rose, carnation and chrysanthemum. Can J Plant Sci 56:525
46. Habson GE, Nichols R (1977) Enzyme changes during petal senescence in carnation. Ann Appl Biol 85:445
47. Schnabl H, Mayer I (1976) Dark fixation of CO_2 by flowers of cut roses. Planta 131:51
48. Weinstein LH, Laurencot HJ (1958) Senescence of rose II. Dark fixation of CO_2 by cut 'Better Times' roses at different stages of senescence. Contrib Boyce Thomp Inst 19:327
49. Coorts GD (1973) Internal metabolic changes in cut flowers. HortScience 8:195
50. Kende H, Hanson AD (1976) Relationship between ethylene evolution and senescence in morning glory flower tissue. Plant Physiol 57:523
51. Hew CS, Thio YC, Wong SY, Chin TY (1978) Rhythmic production of CO_2 by tropical orchid flowers. Physiol Plant 42:226

52. Nichols R (1976) Cell enlargement and accumulation in the gynoecium of the glasshouse carnation (*Dianthus caryophyllus*) induced by ethylene. Planta 130:47
53. Nichols R, Ho LC (1975) Effects of ethylene and sucrose on translocation of dry matter and ^{14}C-sucrose in the cut flower of the glasshouse carnation (*Dianthus caryophyllus*) during senescence. Ann Bot (Lond) 39:286
54. Nichols R, Ho LC (1975) An effect of ethylene on the distribution of ^{14}C-sucrose from the petals to other flower parts in the senescent cut inflorescence of *Dianthus caryophyllus*. Ann Bot (Lond) 39:433
55. Rogers MN (1973) An historical review of postharvest physiology research on cut flowers. Hort Science 8:189
56. Kaltaler REL, Steponkus PL (1976) Factors affecting respiration in cut roses. J Am Soc Hortic Sci 101:352
57. Sacalis JN, Chin CK (1977) Metabolism of sucrose in cut roses. I. Comparison of sucrose pulse and continuous sucrose uptake. J Am Soc Hortie Sci 101:254
58. Nichols R (1975) Senescence and sugar status of cut flower. Acta Hortic 4:21
59. Santarius KA (1973) The protective effect of sugars on chloroplast membrane during temperature and water stress and its relationship to frost, desiccation and heat resistance. Planta 113:105
60. Simpson DJ, Bagar MR, Lee TH (1975) Ultrastucture and carotenoid composition of chromoplast of the sepals of *Sterlitzia reginae* Aiton during floral development. Ann Bot (Lond) 39:175
61. Valadon LRG, Mummery RS (1969) Changes in carotenoid composition of certain roses with ages. Ann Bot (Lond) 33:671
62. Packet RC (1966) Color changes in flowers of *Lathyrus hirsutus* during senescence. Nature 211:12, 15
63. Stead AD, Moore KG (1977) Flower development and senescence in *Digitalis purpurea* L., cv. Foxy. Ann Bot (Lond) 41:283
64. Stickland RG (1972) Changes in anthocyanin, carotenoid, chlorophyll and protein in developing florets of the chrysanthemum. Ann Bot (Lond) 36:459
65. Stewart RN, Norris KH, Asen S (1975) Microspectro-photometric measurement of pH and pH effect on colour petal epidermal cells. Phytochemistry (Oxf) 14:937
66. Asen S, Stewart RN, Norris KH (1975) Anthocyanin flavanol co-pigments and pH responsible for larkspur flower color. Phytochemistry (Oxf) 14:2677
67. Asen S, Norris KH, Stewart RN (1971) Effects of pH and concentration of the anthocyanin flavanol co-pigment complex on the color of 'Better Times' roses. J Am Soc Hortic 96:770
68. Zimmerman PW, Hitchcock AE, Crocker W (1931) The effect of ethylene and illuminating gas on roses. Contrib Boyce Thompson Inst 3:459
69. Hitchcock AE, Crocker W, Zimmerman PW (1932) Effect of illuminating gas on the lily, narcissus, tulip, and hyacinth. Contrib Boyce Thompson Inst 4:155
70. Uota M (1969) Carbondioxide supression of ethylene-induced sleepiness of carnation blossom. J Am Soc Hortic Sci 94:598
71. Barden LE, Hanan JJ (1972) Effect of ethylene on carnation keeping life. J Am Soc Hortic Sci 97:785
72. Kende H, Baumgartner B (1974) Regulation of aging in flowers of *Ipomoea tricolor* by ethylene. Planta 116:279
73. Akamine EK (1963) Ethylene production in fading vanda orchid blossoms. Science 140:1217
74. Arditti J, Hogan NM, Chadwick AV (1973) Postpollination phenomena in orchid flowers. IV. Effect of ethylene. Am J Bot 60:883
75. Burg SP, Dijkman MJ (1967) Ethylene and auxin participation in pollen induced fading of orchid blossoms. Plant Physiol 42:1648
76. Mayak S, Dilley DR (1976) Effect of sucrose on the response of cut carnation to kinetin, ethylene and abscisic acid. J Am Soc Hortic Sci 101:583
77. Mayak S, Dilley DR (1976) Regulation of senescence in carnation (*Dianthus caryophyllus*): Effect of abscisic acid and carbon dioxide on ethylene production. Plant Physiol 58:663

78. Maxie EC, Farnham DS, Mitchel FG, Sommer NF, Parson RA, Snyder RG, Rae HL (1973) Temperature and ethylene effects on cut flowers of carnation (Dianthus caryophyllus). J Am Soc Hortic Sci 98:568
79. Mayak S, Halevy AH, Kats M (1972) Correlative changes in phytohormones in relation to senescence in rose petals. Physiol Plant 27:1
80. Smith WH, Meigh DF, Parker JC (1964) Effect of damage and fungal infection on the production of ethylene by carnations. Nature 204:92
81. Mayak S, Kofranek AM (1976) Altering the sensitivity of carnation flowers (*Dianthus caryophyllus*, L.) to ethylene. J Am Soc Hortic Sci 101:503
82. Mayak S, Kofranek AM, Tirosh T (1978) The effect of inorganic salts on the senescence of *Dianthus caryophyllus* L. flowers. Physiol Plant 43:282
83. Halevy AH (1976) Treatments to improve water balance of cut flowers. Acta Hortic 64:223
84. Nichols R (1977) Sites of ethylene production in the pollinated and unpollinated senescing carnation (*Dianthus caryophyllus*, L.) inflorescence. Planta 135:155
85. Mayak S, Halevy AH (1970) Cytokinin activity in rose petals and its relation to senescence. Plant Physiol 46:497
86. Heide OM, Oydvin J (1969) Effects of 6-benzyl-amino-purine on the keeping quality and respiration of glass house carnation. Hortic Res 9:26
87. Garrod JF, Harris GP (1978) Effects of gibberellic acid on senescence of isolated petals of carnation. Ann Appl Biol 88:309
88. Kelley JD, Schlamp AL (1964) Keeping quality, flower size and flowering response of three varieties of Easter lilies to gibberellic acid. Proc Am Soc Hortic Sci 85:631
89. Gilbart DA, Sink KC (1971) Regulation of endogenous indoleacetic acid and keeping quality of poinsettia. J Am Soc Hortic Sci 96:3
90. Wulster G, Sacalis J, Janes HW, Frenkel C (1981) Senescence in isolated carnation petals; effects of IAA and inhibitors of protein synthesis. HortScience 16:404
91. Mayak S, Halevy AH (1972) Interrelationship of ethylene and abscisic acid in the control of rose petal senescence. Plant Physiol 50:341
92. Borochov A, Tirosh T, Halevy AH (1976) Abscisic acid content of senescing petals on cut rose flowers as affected by sucrose and water stress. Plant Physiol 58:175
93. Ballantyne DJ (1965) Senescence of daffodil (*Narcissus pseudonarcissus* L.) cut flowers treated with benzyladenine and auxin. Nature 205:819
94. Addicott FT (1970) Plant hormones in the control of abscission. Biol Rev 45:485
95. Addicott FT, Lyon JL (1973) Physiological ecolgy of abscission. In. Kozlowsky TT (ed) Shedding of plant parts, Academic Press London pp 85–124
96. Hanischten Cate CH, Ploeg-Voogd JJL. Bruinsma J (1973) Begonia. III. Anatomical pattern of abscission. Acta Bot Nelrl 22:681
97. Simons RK (1973) Anatomical changes in abscission of reproductive structures. In: Kozlowsky TT (ed) Shedding of plant parts Academic Press, (London) p 383
98. Esau K (1965) Plant Anatomy. Wiley, New York
99. Addicott FT (1977) Flower behavior in *Linum lewissii*: some ecological and physiological factors in opening and abscission of petals. Am Micl Nat 97:321
100. Porter NG (1977) The role of abscisic acid in flower abscission of *Lupinus lecteus*. Physiol Plant 40:50
101. Halevy AH, Kofranek AM (1976) The prevention of flower bud and leaf abscission in pot roses during simulated transport. J Am Soc Hortic Sci 101:658
102. Peterson JC, Sacalis JN, Durkin DJ (1980) Alterations in abscisic acid content of *Ficus benjamina* leaves resulting from exposure to water stress and its relationship to leaf abscission. J Am Soc Hortic Sci 105:793

Chapter 3 Carnation

3.1 Introduction

Owing to its excellent keeping quality, wide range of forms, ability to withstand long-distance transport and remarkable ability to rehydrate after continuous shipping, carnation is preferred by growers to roses or chrysanthemums in several flower-exporting countries. According to a study conducted in the USA, approximately 34% of the total number of consumers expressed their liking for carnation 5 days after they were purchased, as against only 20% who favoured roses [1]. Although cut carnations are sold all the year round, they are in particular demand for Valentine's Day, Easter, Mother's Day and Christmas. While standard carnations are in greater demand, the miniature types are fast gaining in popularity for their potential use in floral arrangements and also as a cut flower at comparatively low price. White and pink standard carnations are in greatest demand, followed by red, yellow, sky blue and bicoloured [2]. In contrast, in miniatures, the terminal flowers bud is pinched to encourage clusters of short-stemmed small flowers. These types are mainly used for flower arrangements.

One of the unique features of the American carnation industry is that the major production centres are located in areas with ideal or nearly ideal growing conditions [3]. The carnation flowers are normally air-shipped over long distances.

3.1.1 Postharvest Quality Requirements

Stem length and physical condition of the flower are the two major postharvest considerations in handling cut carnations. Most of the grades presently followed in different countries are mainly based on these two parameters. The Society of American Florists (SAF) suggested the following inspection standards for cut carnations [1]:

1. Bright, clean, firm flowers and leaves.
2. Fairly tight petals near the centre of the flowers that are tight and unopened.
3. Symmetrical flower shape and size characteristic of the variety.
4. No split or mended calyx-flowers with split and mended calyx must not carry SAF emblem.
5. No buds or suckers.
6. No decay or damage.
7. Straight stem and normal growth.

Table 3.1. Grades and standards for carnation followed by (a) The Society of American Florists' [1]

Measuring standards	Grades			
	Blue	Red	Green	White
Minimum flower diam.(cm)	7.0	5.7	No	No
Stem length (cm)	55	42.5 – 55.0	25 – 42.2	Any
Stem strength[a]	10 – 2	10.2	Unrestricted	–

[a] The flower head, will not bend lower than ten or two on a clock if the base of the stem is held upright at 6.0 o'clock

(b) European Economic Community [1]

Code	Stem length (cm)
0	less than 5
5	5 – 10
10	10 – 15
15	15 – 20
20	20 – 30
30	30 – 40
40	40 – 50
50	50 – 60
60	60 – 80
80	80 – 100
100	100 – 120
120	More than 120

The grades and standards developed by the SAF and European Economic Community (EEC) are given in Table 3.1. All of these methods do not take into consideration the potential lasting qualities of cut flowers. Several researchers have employed different qualitative (flower and bud development, retention of flower colour) and quantitative (vase life, storage life, water uptake etc.) parameters for the evaluation of postharvest qualities of cut carnation [6 – 13, 113, 114]. In all of these methods, the vase life and ornamental (aesthetic) value of flowers are determined based on subjective evaluation. Buys and Cours [8] used the cumulative water uptake by individual flowers as a basis for postharvest quality and developed a statistical model for the estimation of postharvest lasting quality in roses. Similar correlative studies should also be undertaken in carnations. Standard procedures for the determination of vase life in different cut flowers have been proposed by Reid and Kofranek [7].

The relative importance of several quality attributes have been studied in cut carnations. The condition of the flower is by far the most important consideration (25 points out of 100) in respect of consumer acceptance of cut flowers in carnation, and is followed by form (20 points), colour and size of flower (15 points). Freedom from infection and price and condition of foliage [13] also influence the overall quality and consumer acceptability in carnations [13].

3.1.2 Commercial Cultivars

The choice of cultivars changes rather rapidly depending on consumer preference, technological progress in respect of production and market outlets. However, in the case of carnation, William Sim and its mutants still maintain their popularity over other types and may continue to be preferred for some time [2]. Most of the cultivars from this group have almost all the desirable characteristics of an ideal cut carnation, but relatively weak stem as compared to other types. Vase life of important carnation cultivars in water with or without silver thiosulfate (STS) is given in Table 3.2. The response of cultivars to STS differs considerably. In general, standard types respond better than spray types. The yielding ability of florist carnation depends mainly on the cultivar and growing conditions. The yielding potential of three leading carnation cultivars is given in Table 3.3. The productivity per unit areas can be increased by harvesting in flowers at earlier stage (III) [17].

Table 3.2. Vase life of various carnation in water with or without silver thiosulfate [112]

Cultivar	Vase life (days)	
	Water	STS^z
Standard types		
Scania	5.8	13.3
White Sim	5.2	13.7
Lena	4.5	12.3
Le Reve	8.0	14.0
Calypso	7.0	17.2
Spray types		
Exquisite	8.1	8.5
Scarlet Elegance	10.2	10.6
Corona	12.5	12.9
Tony	12.3	16.2
Sam's Pride	10.7	12.1
Red Baron	9.3	11.2
Silvery Pink	12.4	14.4
Super Gold	10.9	11.7
May Time	9.9	12.3

^{z} STS = Silver thiosulfate

Table 3.3. Yielding potentials of three florist carnation cultivars [17]

Cultivars	Parameters	Bud stage at harvest		
		III	IV	V
Dusty Sim	No. of fl[a]/pl[b]	3.0	3.0	2.5
	No. of fl/m^2	136.8	133.6	113.0
	Percent of control	121.1	118.1	100.0
	Grade I (%)	57.3	50.3	51.1
	Grade II (%)		31.4	26.4
	Grade III (%)	12.4	16.4	17.6
	Out of grade (%)	1.1	1.9	4.9
Scania	No. of fl/pl	3.1	3.0	2.9
	No. of fl/m^2	140.8	136.5	129.2
	Percent of control	109.8	105.6	100.0
	Grade I (%)	51.3	54.9	42.3
	Grade II (%)	29.0	25.6	26.7
	Grade III (%)	17.0	16.4	23.8
	Out of grade (%)	2.7	3.1	7.2
Alice	No. of fl/pl	3.8	3.7	2.4
	No. of fl/m^2	172.9	164.4	126.6
	Percent of control	162.2	154.2	100.0
	Grade I (%)	43.4	38.1	23.4
	Grade II (%)	28.4	30.3	27.1
	Grade III (%)	24.0	28.1	38.7
	Out of Grade (%)	4.2	3.5	10.8

[a] fl = flowers
[b] pl = plant

3.1.3 Production

The major world production centres for carnation are the USA, the Netherlands, Israel, Colombia and the UK [2, 14–16]. In many cases, the production and sales figures for standard and miniature carnation are grouped together for the census. Since the growth habit and production techniques are different for these types, combining the figures for both the groups does not give a clear picture of their relative contribution to the total sales.

In the USA, the leading production centres for carnation are located in the states of Colorado and California, accounting for approximately 20% and 70% respectively of total production for 1981 [14–16].

3.2 Physiology and Biochemistry

3.2.1 Growth and Development

Temperature and light are the two major factors determining the success of the carnation industry. The optimum range of night temperatures for carnation during winter is 10 to 11 °C, while the same during spring and summer is 12.7 and 13.0 to 15.4 °C, respectively [2]. The optimum growing temperatures will depend upon the available solar radiation [2]. Higher day and night temperatures, especially during flowering, result in abnormal flower opening and calyx spliting [18]. Best quality carnations are produced in areas having high light intensity during winter, when at the same time the temperatures during the summer months are mild.

The colour, grade and quality of cut carnation are influenced by the growing temperatures [18–20]. In the case of the Red Gaiety carnations, best colour, grade and postharvest quality of flower were obtained by maintaining the night temperatures in the greenhouse at 12.1 °C. Day temperatures, on the other hand, also affect the mean grade flowers. A day temperature of 18.2 °C was shown by Hanan [19] to be optimum for obtaining a maximum number of fancy grade carnations. At this temperature the mean flower grade was also higher.

Although carnation was considered as a long day plant in the past [21], present-day carnations are considered to be relatively insensitive to photoperiod. Blake [22] classified carnation as a quantitative long day plant. Generally, long photoperiods promote flowering in carnation while short days tend to delay it. Although the early stages of flower development are favoured by long days, the later stages of bud development are controlled by light intensity [23–28]. The shoots with only four to seven pairs of leaves are more sensitive to light intensity and photoperiods than at other stages.

Cyclic lighting from dusk to dawn increases the flower fresh weight and strength of the flower stalk [28]. At the same time, it also increases the number of lateral vegetative shoots as compared to continuous lighting. However, the flower quality is adversely affected when the plants are maintained continuously under long photoperiods, which causes excessive elongation of internodes [30]. The flower quality can be improved by providing long days for only a short period (4 to 6 weeks) when the shoots have four in seven pairs of leaves [29].

3.2.2 Senescence of Cut Carnations

Senescence in carnation flowers has been studied extensively by several researchers, as it provides a classic example of ethylene action in plant tissues [4, 5, 31–66]. It involves manifold rise in the production of ethylene, in-rolling of corolla, growth of gynoecium and eventual death [54, 55]. Nichols and Ho [44] suggested that all the events leading to flower death are often orchestrated by the growth of the gynoecium, which has been shown to enlarge during senescence. The enlargement of the gynoecium creates competition for the limited source of food supply and because of its strong sink effects, the gynoecium dominates other parts leading to the in-rolling of petals (sleepiness) and eventual death [51–54]. A detailed account on senescence of cut flower is given in Chapter 2.

Mayak et al. [36] summarized the events causing senescence in carnations as follows:

1. An increase in the ethylene level.
2. An equilibrium between low but fairly constant production rate and diffusion from flower. Any factor which increases the internal ethylene production initiates some event having a lag phase of 2 h followed by a long linear phase of production.
3. Increase in tonoplast permeability which takes about 2 h. It destroys the semi-permeability of membranes and osmotic gradient.
4. Decrease in water uptake which closely follows increase in tonoplast permeability.
5. Visual symptoms of wilting occurs about 2 h after the decline in water uptake has began due to loss of turgor.

3.3 Postharvest Losses

3.3.1 Nature of Losses

Losses in carnation result mainly from exposure of flowers to high temperatures, water stress and any other conditions that accelerate ethylene synthesis or render the flower more sensitive to ethylene. Within reasonable limits, wilting of carnation flowers is reversible, but extended exposure to unfavourable environment causes irreversible wilting. The physical condition of the flower is the sole determinant of consumer acceptability in cut carnations. Split or mended calyx, asymmetrical or fully opened flower head, sleepy flowers, presence of buds or suckers at the basal portion of the stalk, weak stems, and damaged flowers and foliage reduce the quality and the postharvest life of flowers at room temperatures. Colour of flowers also influences their marketability. Petal senescence, which is observed in flowers exposed to very high concentrations of ethylene or packed with aged flowers, also causes postharvest losses.

3.3.2 Causes of Losses

Cultural Causes. Growing temperatures, light intensity and CO_2 concentrations in the growing environment influence the postharvest lasting qualities of carnations. Flowers that are grown in areas where the weather during flower bud development is warm are more susceptible to adverse conditions during storage or shipment, that quite often cause sleepiness [48]. Flowers grown at 23.7 °C or higher had shorter postharvest life than those grown at 15.4 °C [2]. However, flower size and stem strength were better when grown at 23.7 °C. CO_2 concentration during the growth cycle of carnation also influences the lasting qualities of many florist carnation cultivars.

Stage of harvesting is another important factor that influences postharvest losses. The buds harvested at very early stages (Stages 1 and 2, Fig. 3.1) do not open properly under laboratory conditions [6]. The flower size and vase life of these flowers were also considerably reduced [17] (Table 3.3). The buds harvested at too tight a stage usually took longer to open than those harvested

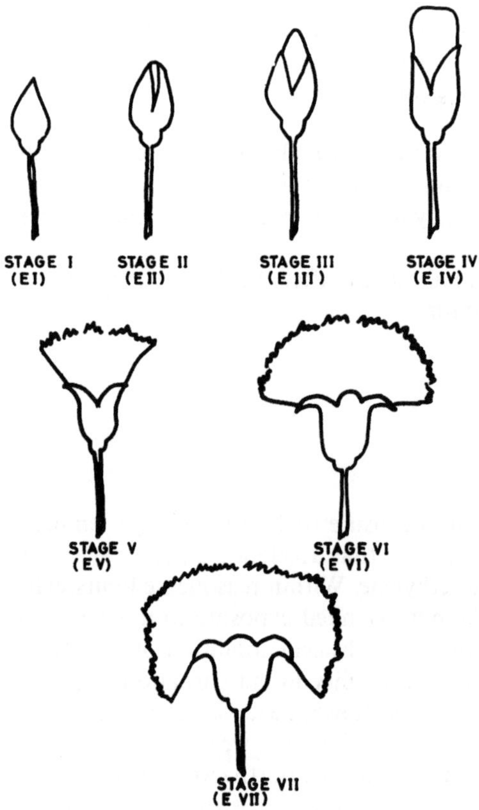

Fig. 3.1. Stages of flower bud development in carnation [6]

at later stages of development. Conditions during bud opening and in holding areas such as light, temperature, and relative humidity also have an impact on the subsequent vase life [1].

Mechanical Causes. Bruising and physical damage to flowers and stems in transit accelerate the deteriorative processes leading to senescence of petals. Exposure of cut flowers to sunlight or high temperatures at any time during shipment causes sleepiness in carnation [48] (Fig. 3.2). The incidence of sleepiness was much higher, especially when the flowers were kept in cold storage for some time and then brought out for grading and packing, during which the flowers were warmed. Length and temperature of storage also has adverse effects on vase life. Longer storage at comparatively higher temperatures reduces the post-storage vase life at room temperature.

Physiological Disorders. Sleepiness is the single factor causing huge postharvest losses in cut carnations. It can be either temporary (reversible) or permanent (irreversible). Exposure of flowers to ethylene gas or water stress causes this disorder [72]. The incidence of sleepiness is much higher when

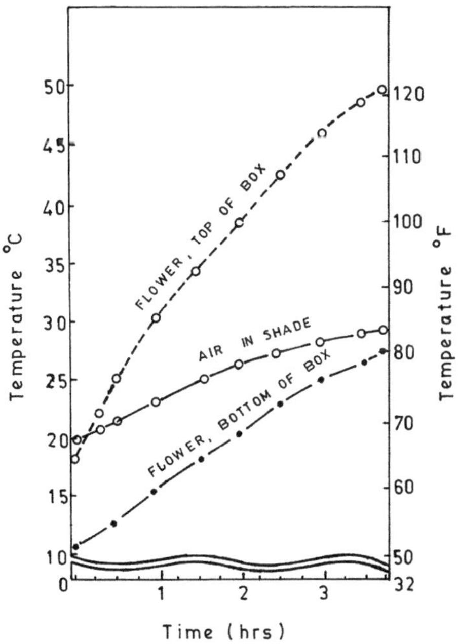

Fig. 3.2. Effects of 4-h exposure of packed containers of Improved White Sim carnations to direct sun light [48]

the flowers are stored for a longer period or are exposed to high temperatures [71].

Nutrient deficiencies also result in postharvest loss due to reduced quality of the product. Deficiency of potassium has been shown to reduce water uptake, tolerance to ethylene and potential life of flowers [36]. Adequate amounts of inorganic salts like potassium chloride, ammonium and potassium nitrate delayed senescence by maintaining the osmotic potential in petal tissues and substituted metabolic sugars in extending vase life of carnation [88].

Postharvest losses due to wastage and reduction in quality increase by shipping or storing carnation flowers with fruits and vegetables that produce ethylene (Table 3.4) [67].

Calyx splitting and uneven opening of flowers are very common disorders in areas of warmer climates. These disorders have been associated with several factors viz. high temperature, sudden fluctuations in day and night temperatures, nutritional deficiencies etc. [115, 116]. All of these conditions reduce the marketability of flowers.

Diseases and Pests. Botrytis blight is the most serious among carnation diseases, and causes water-soaked areas that turn grey due to mould growth. It is often serious in white-flowered varieties. The fungus can thrive well at a wide range of temperatures and high humidity. It produces ethylene and thereby accentuates the seriousness of the problem [2].

Red spiders, aphids, and thrips are the important insect pests that attack carnations. Thrips cause streaks in flowers and are often common in warm

Table 3.4. Postharvest life of cut carnation flowers stored with fruits and vegetables at 23.7 °C [67]

Fruit or vegetable stored with carnation	Postharvest life (days)
Harmful	
Apple	0.5
Pear	+0.5
Avocado	1.0
Onion (green)	1.0
Tomato	1.0
Banana	1.5
Pineapple	2.0
Radish (with tops)	2.0
Celery	+2.0
Broccoli	3.0
Bell pepper	−4.0
Lime	4.0
Squash (whole)	4.0
Lemon	5.0
Onion (dry)	−6.0
Orange	6.0
Control	+7.0
Not harmful	
Cabbage	7.0
Cucumber	7.0
Sweet potato	8.0
Carrot	8.0
Potato	8.0
Lettuce	8.0
Grape	9.0
Grape fruit	9.0
Cauliflower	10.0

weather. These insects, apart from reducing the visual appeal of the product, increase ethylene production by wounding the tissues.

3.4 Loss Reduction Biotechnology

The major postharvest loss reduction techniques in carnation include regulation of preharvest growing conditions (nutrition, regulation of light and temperature, CO_2 enrichment, watering, relative humidity control etc.), use of improved harvesting, grading and packaging techniques, use of low temperature storage, low pressure storage, controlled and modified atmospheric storage, use of growth regulators and floral preservatives.

3.4.1 *Cultural Practices*

Temperature management is one of the most important considerations in carnation production. Most optimum temperature ranges depend upon the solar energy available at different times of the year. Both day and night temperatures are important in producing high quality carnation flowers. Decreasing night temperatures can increase the instances of calyx splitting. Six to 13 times more splitting has been reported in plants grown at 4.4 °C than in those produced at 10 °C. Plants produced at 15.4 °C had the lowest number of flowers with split calyx [116]. The optimum temperature ranges for carnation are 10 to 15 °C during winter, 12.7 °C during spring and 13.0 to 15.4 °C during summer [29].

The sensitivity of flowers to ethylene can be reduced by acclimatizing the plants to water stress during the early stages of growth and development of flowers [47]. Nutrient deficiencies, especially of calcium and potassium, significantly reduce the postharvest life in cut carnations [1]. Therefore, adequate supply of these nutrients during the growth of the plant is necessary in order to ensure quality flowers.

3.4.2 *Harvesting, Grading and Packaging*

The stage of development of the bud plays an important role in determining success with opening and subsequent postharvest qualities. Bud size and petal growth are used to judge the stage of harvesting. Different stages of flower bud development in carnation are given in Fig. 3.1. One of the unique qualities of carnation is their ability to open under in vitro conditions without any reduction in the postharvest quality of the flowers. Nowak et al. [17] showed that the buds should be harvested after they reached either the large bud (tight buds, size 15 mm diam.) or cross bud stage (when petals are visible, size 15–20 mm). Buds harvested too early did not open properly and had shorter vase life. The total yield was also increased by cutting the flowers in cross or large bud stages as compared to the paint brush stage (petals enlarged, size 20 to 25 mm).

The term "maturity" refers to the point in time when flowers can be harvested, irrespective of their physiological maturity, and it will continue to grow and develop to a maximum quality. The point in time at which flowers should be harvested depends on a number of factors, including stage of maturity, time of day, distance to market, customer demand and time of the year. In general, carnation flowers harvested in bud stage (calyx opening) last 4 to 5 days longer than when they were cut at fully open stage [1]. However, the major disadvantages of bud harvesting are grading problems, uneven opening, failure to open and infections.

The length of stem cut with the flower depends on the time of the year. During fall, the cut should be made high enough to encourage more lateral shoot growth, while the flowers can be cut with longer stems when the plants are to be discarded. The best place for cutting the stem is the area where leaves are well spaced and where at least two axillary shoots appear.

Grades recommended by the SAF and EEC are given in Table 3.1. Following a uniform grading system helps the industry to reduce shrinkage.

Carnations are packed in corrugated cardboard containers. Delays between cutting and packing may sometimes be several hours and, as a result, flowers are exposed to high temperatures. These conditions increase the changes of sleepiness. The boxes or containers should be well insulated and should have a lining of vapour barrier such as polyethylene to help maintain high relative humidities inside the package. The bunches of 25 flowers are then placed in these boxes by hand with one half of the total number of bunches oriented on each end of the container. Close packing of flowers helps to minimize the distance for heat exchange between flowers and container-ambient air interface. Newspaper layers are placed between the layers to maintain high humidity and then, when the container is filled, an insulating layer of paper is put across the box to cover the flowers completely. A strip of wood is inserted laterally in the middle of the box, forced downwards and stapled into place to prevent damage to stems during shipping. The lids are fixed to the box by taping or twining them to the box. Sometimes the top lid is made partially or completely transparent for viewing the products without disturbing them. The containers should always be kept under refrigerated conditions.

3.4.3 Low Temperature Storage

Cooling of flowers before shipping is an absolute necessity in order to prevent sleepiness in cut carnations. Regardless of the stage of bud at harvest, they should be cooled rapidly at −0.6° to 0.0 °C [19]. The relative humidity in the storage environment influences the subsequent keeping quality of flowers. Flowers stored at 27 °C (room temperature) and 98 +% relative humidity lasted the same as those stored at 10 °C and 15% relative humidity. The longest post storage shelf life (113 days) was obtained by maintaining storage room at 10 °C and 98+% humidity [1]. In carnations, the low temperature drastically reduces the rate of respiration [113] (Table 3.5) and sensitivity of flowers to ethylene [48] (Table 3.6).

Under ideal conditions (0 °C and 90+ % humidity and vertical positioning of flowers), cut carnations can be stored for 6 to 8 weeks, but under normal

Table 3.5. Respiratory rate of Improved White Sim carnations stored at 0 to 50 °C [48]

Storage temperature (°C)	Respiratory rate in terms of mg CO_2/h/kg
0	9.7
10	30.0
20	239.0
30	516.0
40	1053.0
50	1606.0

Table 3.6. Concentrations of ethylene and time of exposure needed for unacceptable reduction in vase life of Improved White Sim carnation at various temperatures [48]

Storage temperature (°C)	Exposure time	Ethylene concentration needed (ppm)
0	3 days	22.0
0	4 days	2.60
5	2 days	0.90
5	4 days	0.12
20	6 h	2.78
20	18 h	0.96
30	12 h	0.37
40	5 h	0.37

commercial conditions, they are stored for only 2 weeks [1]. Temperatures below optimum will result in chilling damage, while temperatures above optimum and a short delay in cooling reduce vase life. For example, flowers stored at 3 °C above optimum (−0.6 °C) had 80% shorter vase life than those stored at optimum temperature [1]. Rapid precooling of flowers in package helps to reduce the water loss and prevent sleepiness. The greenhouse temperature during the last week before harvest has a profound influence on vase life after dry storage of flowers. Flowers from a greenhouse with a temperature of 26.4 °C during the last week before cutting were completely spoiled during storage at −0.6 °C for 37 days, while the unstored from the same lot had a shelf life of 4 days [29].

Preconditioning of flowers with silver thiosulfhate or sucrose before dry storage increases their storageability and post-storage vase life [1, 76]. The containers should at no time be removed from cold storage during storage or shipping.

3.4.4 Controlled Atmosphere (CA) Storage

Carbon dioxide acts as a competitive inhibitor in ethylene action [69]. Partial pressures of CO_2 and O_2 influence the production and activity of ethylene around the flowers. High levels of CO_2 markedly suppress the ethylene damage to plant tissue, but high levels of O_2, on the other hand, enhance it [70]. Similar results were reported by several other investigators [38, 39, 58, 68]. Controlled atmospheric storage in carnation involves (1) increasing the levels of CO_2, (2) decreasing the level of O_2, (3) low storage temperature, and (4) prevention of build-up of endogenous ethylene levels to threshold that could trigger senescence reaction [70, 71]. Oxygen level below 0.5% to 1% stimulated anaerobic respiration [70]. Smith and Parker [38] found that CO_2 levels of 0.35% or higher were enough to offset the effects of ethylene at or slightly above threshold levels for injury on carnation (0.05 ppm for 2 to 4 days), while

Table 3.7. Vase life (days) and visual acceptablility (%) of Improved White Sim flowers immediately on removal from different types of storage [76]

Treatments	Storage duration (weeks)			
	Four		Eight	
	Vase life (days)	Visual acceptability (%)	Vase life (days)	Visual acceptability (%)
No storage	12.3	100	12.3	100
Normal	9.3	100	5.7	100
Low pressure	7.6	100	6.6	100
Low oxygen				
0.5% O_2	8.8	100	7.1	100
4% O_2	9.6	100	6.9	100
8% O_2	8.8	100	6.6	100

Storage temperature $-2.9 \pm 1.1\,°C$

Uota and Garazsi [73] found that 7% to 20% CO_2 were effective in suppressing sleepiness in carnation caused by 0.25 ppm ethylene for 24 h at 20 °C. Hanan [70] demonstrated that CO_2 concentrations above 4% were harmful to carnation. Although the beneficial effects of high CO_2 storage are well known, CA storage is presently not economically feasible in the case of carnations [1, 74, 75]. However, Staby et al. [76] stored cut carnations for 8 weeks at low O_2 concentrations (0.5% to 8%) without any adverse effects (Table 3.7).

3.4.5 *Modified Atmospheric (MA) Storage*

The room temperature vase life of cut carnations was extended by storing the flowers under modified atmospheric (MA) conditions created by sealing the flowers in plastic bags. The success of MA storage depends largely upon the permeability of films to CO_2 and O_2, temperature, length of storage, and cultivar [71, 111].

3.4.6 *Low Pressure Storage*

The advantages of hypobaric storage of carnation are well established [1, 76–80]. Bud carnations can be stored for 90+ days under low pressure without losing any vase life or other postharvest qualities [1]. Table 3.7 presents the comparison between different types of storage in cut Improved White Sim cannation [76]. Pretreatment of flowers with silver thiosulfate before storage greatly improved the vase life after removal [76].

3.4.7 *Use of Growth Regulators, Floral Preservative and Bud Opening Solution*

Use of certain growth regulators and preservative solutions is beneficial at all stages of marketing of cut carnation. Hundreds of compounds have been tested for their efficacy in extending the shelf life of cut flowers [41, 45, 81–110]. The ideal floral preservative should maintain water absorption and full turgidity in flower tissues. Among hundreds of chemicals tested only few have the properties for an ideal preservative solution. Soaking the flower stems in 1200 ppm of silver nitrate for 5–20 min improved flower longevity by over 50% [97]. The shelf life of cut carnation was extended from 16 to 22 days by placing the stems in a commercial floral preservative solution prepared with water into which silver ions were added electrolytically [12, 29] (Table 3.8).

Microbial growth in preservative solution can be checked by including 200 to 1000 ppm of 8-HQC, but at higher concentration 8-HQ causes petal injuries in many cut flowers. Addition of sodium benzoate (100 ppm) controlled slime production in the vase solution and improved the effectiveness of 8-HQC. Sodium dichloro-S-triazine-trione (SDT) (250 ppm) plus 3% sucrose extended the vase life from 6 to 8 days to 16 to 20 days without any adverse effects [98, 99]. Of all the compounds tested either alone or in combination, a solution containing 200 ppm 8-HQC+ 3% sucrose; 400 ppm 8-HQC+ 500 ppm B-Nine (not registered) +5% sucrose; or 10% sucrose+2 ppm silver nitrate (not registered) were found to be most useful [1]. Deionized water or water containing less than 600 ppm soluble salts should be used for preparing the preservative solution.

Preshimpment treatment of flowers with silver nitrate (1200 ppm) for 10 min is very beneficial in improving the vase life and storageability [1]. Other preconditioning treatments like dipping entire flower stems in Benlate (4 oz/gallon) or overnight stem dip in 10% sucrose +200 ppm 8-HQC were also found useful in this regard [1].

The advantages of bud harvest in carnation are well established [89]. The success with bud harvest greatly depends on the ability of the bud-cut flowers

Table 3.8. Effects of floral preservative and silver ions on vase life of cut carnations [29]

Preservative solution	Postharvest lasting life (days)
Water	6.3
Water + silver ion	9.5
Flora life[a]	12.9
Flora life[a] + silver ion	15.4
Keep[a]	11.0
Keep[a] + silver ion	20.7
Petal life[a]	14.6
Petal life[a] + silver ion	22.2

[a] Trade name

Table 3.9. Composition of bud-opening solutions [93]

Name of the opening solution	Composition	Reference
Modified cornell	8-hydroxy quinoline sulfate (200 ppm) Silver nitrate (25 ppm) Aluminium sulfate (50 ppm) Sucrose (10%)	[89]
UC solution (Davis)	Silver nitrate (25 ppm) Citric acid (75 ppm) Sucrose (10%)	[89]
Modified Kagawa	Alar (700 ppm) 8-hydroxy quinoline sulfate (400 ppm) Sucrose (6%)	[85] [106]
TOG	8-hydroxy quinoline glycolate (300 ppm) Thiabendazole glycolate (300 ppm) Sucrose (10%)	[81] [82]
BAC[a]	Benzylkonium chloride (500 ppm) Citric acid (75 ppm) Sucrose (10%)	[93]

[a] Contains dichlor-benzyl-dimethyl-alkyl ammonium chloride, (BAC) compound

to open to their normal size under laboratory conditions. Chemical bud-opening solutions are necessary in bud opening because the flowers harvested in the tight bud stage fail to open to their normal size in water. All the known flower bud-opening solutions contain ingredients like water, sugars, germicides, growth regulators and other products [6, 84, 93, 107]. Many of the bud-opening solutions have been used in carnations and their efficacy has been completely described. The opening of buds is greatly improved by providing 150 ft candles of light during the opening [1]. The optimum temperature during opening of floral buds has been found to be around 18.2 °C [1]. The composition of the commonly used bud-opening solutions is given in Table 3.9.

3.4.8 Control of Pests and Diseases

Control of pests and diseases forms an integral part of all the improved technologies in order to reduce shrinkage and increase potential-lasting qualities. Unless the diseases and insect pests are controlled in time in the growing environment, it becomes difficult to control them in the postharvest stage.

References

1. Staby GL, Robertson JL, Kiplinger DC, Conover CA (1978) Chain of Life. Ohio Florists Assoc, Ohio State Univ, Columbus
2. Laurie A, Kiplinger DC, Nelson KS (1968) Commercial flower forcingg. McGraw-Hill Book, New York
3. Ball V (1976) The Ball Red Book. Ball Seed Company. West Chicago M Ill, USA

4. Nichols R (1966) Ethylene production during senescence of flowers. J Hort Sci 91:279
5. Nichols R (1971) Induction of flower senescence and gynoecium development in the carnation (*Dianthus caryophyllus*) by ethylene and 2-chloroethylphosphonic acid. J Hort Sci 323
6. Casp AM, Salvador PJ, Ibanenez MJ (1980) Objective characterization of stages in bud opening of carnation (*Dianthus caryophyllus* L.). Acta Hortic 113:175
7. Reid MS, Kofranek AM (1980) Recommendations for standardized vase-life evaluation. Acta Hortic 113:171
8. Buys CA, Cours HG (1980) Water uptake as a criterion for vase life of cut flowers. Acta Hortic 113:127
9. Salinger JP (1975) Criteria for evaluation of postharvest senescence of cut flowers. Acta Hortic 41:201
10. Systema W (1975) Conditions for measuring the vase-life of cut flowers. Acta Hortic 41:217
11. Sass JE (1968) Botanical microtechnique. The Iowa State Univ Press, Ames, Iowa, p 228
12. Halevy AH, Kofranek AM (1977) Silver thiosulphate of carnation flowers for reducing ethylene damage and extending longevity. J Amer Soc Hort Sci 102:76
13. McDaniel GL (1982) Ornamental Horticulture. Reston, USA
14. Anon (1980) Floriculture Crops Production Area and Sales. Statistical Reporting Service, USDA, Washington DC
15. Anon (1981) Floriculture Crops – Production, Area and Sales. Statistical Reporting Service, USDA, Washington DC
16. Anon (1982) Floriculture Crops – Production Area and Sales. Statistical Reporting Service, USDA, Washington DC
17. Nowak J, Mynett K, Rudhicki RM (1983) The effect of cutting carnations in the bud stage on quality and yield of flowers from a nursery. Sci Hort 20:203
18. Schmidt RG, Holley WD (1957) Some effects of night temperature on carnations. Colo Flower Grow Assoc Bul 93:1
19. Hanan JJ (1959) Influence of day temperatures on growth and flowering of carnations. Proc Amer Soc Hort Sci 74:692
20. Hanan JJ (1969) Temperatures in carnation flowers. Colo Flower Grow Assoc Bul 226:1
21. Withrow RB, Richmon MW (1933) Artificial radiation as means of forcing greenhouse crops. Purdue Univ Agric Exp Sta Bul 380
22. Blake J (1955) Photoperiodism in perpetual flowering carnation. Rept 14th Int Hort Congress 1:331
23. Blake J, Spencer R (1982) The influence of temperature during the light and dark period on the growth and flowering of carnation. Rept 15th Int Hort Congress 412
24. Harris GP, Ashford M (1966) Promotion of flower initiation in the glasshouse carnation by continuous light. J Hort Sci 41:397
25. Harris GP, Bradber AP (1966) The flowering behavior of carnation cuttings produced during long and short days. J Hort Sci 41–73
26. Harris GP (1972) Intermittant illumination and photoperiodic control of flowering incarnation. Ann Bot 36:345
27. Harris GP, Harris JE (1962) Effects of environment on flower initiation in carnation. J Hort Sci 37:219
28. Metlerz B (1972) Cyclic lighting of carnations. Colo Flower Growers Assoc Bul 265:1
29. Mastalerz JW (1978) Greenhouse Environment. Wiley, New York USA
30. Cox RJ (1967) Controlled flowering of carnation by dusk to dawn lighting. Sparkes Tech Rept AG, Sparks, Sussex UK
31. Nichols R (1973) Senescence of carnation flower – Respiration and sugar status. J Hort Sci 48:111
32. Fujino DW, Reid MS, Yang IF (1980) Effects of aminooxyacetic acid on postharvest characteristics of carnation. Acta Hortic 113:59
33. Jona R, Accati E, Mayak S (1980) Senescence processes as reflected in change in polysaccharidic cell wall components. Acta Hortic 113:153
34. Camprubi P, Nichols R (1978) Effects of ethylene on carnation flowers (*Dianthus caryophyllus*) cut at different stages of development. J Hort Sci 53:17

35. Mor Y, Reid MS (1980) Isolated petals – A useful system for studying flower senescence. Acta Hortic 113:19
36. Mayak S, Kofranek AM, Tirosh T (1978) The effects of inorganic salts on senescence of *Dianthus caryophyllus* flowers. Physiol Plant 59:591
37. Crocker W, Knight LI (1908) Effects of illuminating gas and ethylene on flowering carnations. Bot Gaz 46:259
38. Smith WH, Parker JC (1966) Prevention of ethylene injury to carnation by low concentrations of carbon dioxide. Nature 211:100
39. Uota M (1969) Carbon dioxide suppression of ethylene induced sleepiness in carnation blooms. J Amer Soc Hort Sci 54:598
40. Dilley DR, Carpenter JW (1975) The role of chemical adjugants and ethylene synthesis on cut flower longevity. Acta Hortic 41:117
41. Eisinger W (1977) Role of cytokinins in carnation flower senescence. Plant Physiol 59(4):707
42. Nichols R (1980) Development in postharvest techniques for cut flowers. Sci Hort 31:43
43. Borden LE, Hanan JJ (1972) Effect of ethylene on carnation keeping life. J Amer Soc Hort Sci 89:677
44. Nichols R, Ho LC (1975 a) Effects of ethylene and sucrose on translocation of dry matter and ^{14}C sucrose in the cut flower of the glasshouse carnation (*Dianthus caryophyllus*) during senescence. Ann Bot 39:287
45. Mayak S, Dilley DR (1976) Effects of sucrose on response of cut carnation to kinetin, ethylene and abscisic acid. J Amer Soc Hort Sci 101:583
46. Acock B, Nichols R (1979) Effects of sucrose on water relations of cut senescing carnation flowers. Ann Bot 44(2):221
47. Mayak S, Kofranek AM (1976) Altering the sensitivity of carnation flowers of *Dianthus caryophyllus* to ethylene. J Amer Soc Hort Sci 101:503
48. Maxie EC, Farnham DS, Mitchell F, Sommer NF, Parsons RA, Snyder RG, Rae HL (1973) Temperature and ethylene effects on cut flowers of carnations (*Dianthus caryophyllus* L.). J Amer Soc Hort Sci 98:568
49. Nichols R (1977) A descriptive model of the senescence of carnation (*Dianthus caryophyllus*) inflorescence. Acta Hortic 71:227
50. Nichols R (1970) Senescence of carnations metabolism of sugars by cut flowers and ethylene accumulation in petals. Glasshouse Crops Res Inst Ann Rep 73
51. Mor Y, Reid MS, Kofranek AM (1980) Role of ovary in carnation senescence. Sci Hortic (Amst) 13:377
52. Halevy AH, Mayak S (1979) Senescence and postharvest physiology of cut flowers. Part I. Hortic Rev 1:204
53. Mayak H, Halevy AH (1980) Flower senescence. In: Thimann KV (ed) Plant Senescence. Senescence in Plants, CRC Bocca Raton USA, 131
54. Halevy AH, Mayak S (1981) Senescence and postharvest physiology of cut flowers. Part II. Hortic Rev vol 3 AVI, USA
55. Nichols R (1977) Sites of ethylene production in pollinated and unpollinated senescing carnation (*Dianthus caryophyllus*) inflorescence. Planta 135:155
56. Cameron AC, Fenton CAE, Yu Y, Adams DO, Yang SF (1979) Increased production of ethylene by plant tissues treated with L-aminocyclopropene-L. carboxylic acid. HortScience 14:178
57. Buffler G, Mor Y, Reid MS, Yung SF (1981) Changes in L-aminocyclopropane-L-carboxylic acid contents of cut carnation flowers in relation to their senescence. Planta 150(5):439
58. Nichols R (1981) Ethylene – present and future. Acta Hortic 113:11
59. Larsen FE, Frolish M (1969) The influence of 8-hydroxyquinoline citrate, N-dimethyl aminosuccinamic acid and sucrose on respiration and water flow in Red Sim cut carnation in relation to flower senescence. J Amer Soc Hort Sci 94:289
60. Mayak S, Vaadi Y, Dilley DR (1977) Regulation of senescence in carnation (*Dianthus caryophyllus*) by ethylene-mode of action. Plant Physiol 59:591
61. Nichols R (1976) Cell enlargement and sugar accumulation in gynoecium of the glasshouse carnation (*Dianthus caryophyllus* L.) induced by ethylene. Planta 130:47

62. Nichols R, Ho LC (1975) An effect of ethylene on the distribution of ^{14}C sucrose from the petals to the other flower parts in the senescent inflorescence of *Dianthus caryophyllus*. Ann Bot 39:433
63. Ho LC, Nichols R (1975) Role of phloem transport in the translocation of sucrose along the stem of carnation cut flowers. Ann Bot 39:439
64. Garrud SF, Aurris GP (1978) Effect of gibberellic acid on senescence of isolated petals of carnation. Ann Appl Biol 88(2):309
65. Reid MS (1980) Postharvest handling of cut flowers. Univ Calif Co-op Extn Serv Sept pp 1–4
66. Jeffcot B, Harris GP (1972) Hormonal regulation of the distribution of ^{14}C labelled assimilates in flowering shoots of carnation. Ann Bot 36:353
67. Wintz RP (1954) Storage of cut carnation with some fruits and vegetables. Colo Flower Grow Assoc Bul 53:1
68. Smith WH, Freeman WW (1966) Exposure of cut flowers to ethylene in the presence and absence of carbon dioxide. Nature 211:99
69. Burg SP, Thimann KV (1954) The physiology of ethylene formation in apples. Proc Natl Acad Sci 45:335
70. Hanan JJ (1967) Experiments with controlled atmosphere storage of carnations. Proc Amer Soc Hort Sci 90:370
71. Hardenburg RE, Uota M, Parson SS (1969) Refrigeration and modified atmosphere storage for improving keeping quality of cut flowers. Proc 12th Inst Cong Refrig (Madrid) 3:339
72. Uota M (1970) Sleepiness of carnation blooms – how much ethylene does it take. Flori Rev 33:66
73. Uota M, Garazsi M (1967) Quality and display life of carnation blooms after storage in controlled atmosphere. USDA Marketing Res Rep 796:9
74. Tinga JH (1956) The effect of modified atmosphere at low temperature and treatments after low temperature storage which affect the keeping quality of cut flowers. Ph D Dissertation, Cornell Univ, pp 189
75. Hanan J (1966) Controlled atmospheric storage of carnations. Colo Flower Grow Assoc Bul 193:1
76. Staby GL, Cunningham MS, Holstead CL, Kelly JW, Konjoian PS, Eisenberg BA, Dressler B (1983) Storage of roses and carnation flowers. J Amer Soc Hortic Sci 109:193
77. Lougeed EC, Murr DP, Berard L (1978) Low pressure storage of horticultural crops. HortScience 13:21
78. Dilley DR (1977) The hypobaric concept for controlled atmospheric storage. Michigan State Univ Hort Rept 28:29
79. Dilley DR, Carpenter WJ, Burg SP (1975) Principles and application of hypobaric storage of cut flowers. Acta Hortic 41:249
80. Burg SP (1973) Hypobaric storage of cut flowers. HortScience 8(3):202
81. Apelbaum A, Katchansky M (1977) Improving the quality and prolonging the vase life of bud cut flowers by pretreatment with thiabendazole. J Amer Soc Hort Sci 102(5):623
82. Carpenter WD, Dilley DR (1973) A benzothiadiazole extends carnation flower life. HortScience 8(4):335
83. Casp AM, Escrira I (1978) Un unevoproducto para prolongar la vida de los cleveles cortedos. Vergel Iv (2):5
84. Cywinska-Smoter K, Rudnikii RM, Gozziynska D (1978) The effects of exogenous growth regulators in opening tight carnation buds. Sci Hortic (Amst) 9:155
85. Hasegawa A, Manabe M (1976) Studies on the keeping quality of cut flowers. II. On the floral preservative Kagawa solution and its practical use for cut carnations. Kagawa Daigaku Nogakubu Kokuku 27:59
86. Holley WD, Cheng LH (1967) Harvesting and handling of immature carnations. Proc Amer Soc Hortic Sci 90:377
87. Farnham DS, Ubeda T, Kofranek AM, Mclain AH, Kabota J (1978) Physan-20 an effective biocide for conditioning and bud opening of carnation. Flor Rev 162(4190):24, 58
88. Halevy AH (1976) Treatments to improve water balance of cut flowers. Acta Hortic 64:223
89. Kofranek AM, Farnham DS, Maxie EC, Kabota J (1972) Long term storage of carnation buds. Flor Rev 151(3906):29

90. Larsen FE, Scholes JF (1966) Vase life and flower size of carnation improved with chemicals. Flor Rev 137:29
91. Maronsky FJ (1976) Control of bacteria in cut flower vase water. Proc Fla State Hortic Soc 90:294
92. Purups EV, Chan P (1975) Extension of vase life of cut flowers by use of isoascorbate containing floral preservative solutions. J Amer Soc Hortic Sci 98(1):22
93. Casp AM, Salvador PJ, Ibanez MJ (1981) A new solution for carnation bud opening with promising improvements due to quaternary ammonium compound. Acta Hortic 113:183
94. Heide DM, Oydoin J (1969) The effects of 6-benzylaminopurine on keeping quality and respiration of glasshouse carnation. Hortic Res 9:26
95. Aarts JFT (1957) Over de houdbaarheid snijbloemen Meded Land noucohogesthool, Wageningen 57(9):1
96. Weinstein LH, Laurencot HJ Jr (1963) Studies on the preservation of cut flowers. Contrib Boyce Thompson Inst 22:81
97. Kofranak AM, Paul JL (1972) Silver impregnated stems aid carnation flower longevity. Flor Rev 15(3913):24
98. Kofranek AM, Kohl HC, Kabota J (1974) A slow release chlorine compound as a vase water additive. Flor Rev 154(4000):63
99. Kofranek AM, Kohl HC, Kabota J (1975) Stabilized chlorine compound as a vase water additive. Calif Flower Nursery Rep Jan-Feb: 5–7
100. Kohl HC Jr, Smith DG (1960) Development of carnation flowers cut before fully open. Carnation Craft 53:7
101. Parups EV (1975) Inhibition of ethylene synthesis by benzylisothiocyanate and its use in delaying senescence of cut carnation. HortScience 10(3):221
102. Kofranek AM, Kabota J (1972) Bud opening solutions for carnations. Flor Rev (3905):30
103. Apelbaum A, Katchansky M (1978) Effects of thiabendazole on ethylene production and sensitivity to ethylene of bud cut flowers. HortScience 13:593
104. Farnham DS, Barr C Jr (1973) Carnation buds offer exciting prospects for merchantizers and consumers. Calif Agr Extn Flower Nurs Rept Sept 173:1
105. Hardenberg RE, Vaught HC, Brown GA (1970) Development and vase life of bud cut Colorado and California carnations in preservative solutions following air shipment to Maryland. J Amer Soc Hortic Sci 95:18
106. Hasegawa A, Manabe M, Goi M, Ihara Y (1977) Studies on the keeping quality of cut flowers. II. On the floral preservative "Kagawa Solution" and its practical use for cut carnation. Hortic Abstr 47(3):No 2758
107. Farnham DS, Burr C, Halevy AH (1971) The value of using chemical solutions for conditioning and bud opening carnations. Flor Rev 148(3846):27
108. Le Masson B, Nowak J (1981) Cut flower life of dry transported carnation as influenced by different silver form pretreatments. Sci Hortic (Amst) 15:383
109. Reid MS, Paul JL, Farhoomand MB, Kofranek AM, Staby GL (1980) Pulse treatments with silver thiosulfate complex extend the vase-life of cut carnations. J Amer Soc Hortic Sci 105:25
110. Veen H (1979) Effects of silver on ethylene synthesis and action in cut carnation. Planta 145:467
111. Haugue AW, Bryant Y, Laurie A (1947) Prepackaging of cut flowers. Proc Amer Soc Hortic Sci 49:427
112. Systema W (1980) Vase life and development of carnation as influenced by silver thiosulfate. Acta Hortic 113:33
113. Systema W (1979) Voorbehandeling Van angers. Proefstation voor de bloemisterij in Nederland – Aalmeer. Publication No 6
114. Slootman JE, Molenaar WH (1979) Test van een houdbaarheidsmiddel voor alle snijbloemen. Sprenger Instituut, Wageningen, Publication No 2045
115. Szendel A (1938) The effect of temperature on splitting of carnations. Proc Amer Soc Hortic Sci 36:760
116. Holiday WG, Watson DP (1953) Influence of temperature on flowering and calyx splitting of greenhouse carnations. Proc Amer Soc Hortic Sci 36:538

Chapter 4 Rose

4.1 Introduction

Roses, owing to their wide variety in flower colour, form, growth habit and plant size, have widespread consumer acceptance. Cut roses are generally preferred (51%) to carnations (41%) and chrysanthemums (4.5%) when they are fresh [1]. They account for approximately 39% of the total sale of fresh flowers in the United States of America [2, 11]. Roses are in great demand for Valentine's Day, Sweetheart's Day, Memorial Day and Christmas. Furthermore cut roses are commonly used for many religious and social events such as marriages, birthdays, and social gatherings and are regularly exchanged between loved ones.

All the present-day varieties have been developed from wild rose species [3, 4]. Two kinds of roses are commercially grown for cut flowers. The Hybrid Tea (HT) roses form the bulk (about 90% of the total area under rose) of cut roses that are sold in the major markets of the USA [8]. These cultivars produce a large terminal flower bud and several small lateral buds. The lateral buds are removed to encourage the growth of terminal buds. The demand for HT roses depends upon colour, sex and season of the year. Generally, red HT's are in greatest demand, followed by yellow, pink and white [8].

The second group, Floribundas, which are sometimes referred to as "miniature" roses, form a terminal bud and several axillary buds that are as vigorous and tall as the terminal ones. In cultivars belonging to this group, the terminal or apical flower bud is removed to encourage the development of axillary or lateral buds. Pink Floribundas are preferred by the public to red, yellow, white or bicolour types. The stem length in these types is not a major quality consideration. They make beautiful flower arrangements. The Floribundas contribute nearly 10% of the total area under rose, but these types could account for more than 10% of the total sale. In recent years, short-stemmed roses (40 cm or less) and smaller flower types, Sweethearts, are capturing the market, and within a few years they may constitute the largest portion of the roses types sold in the world.

Some of the wild species of rose (*Rosa bourboriana* and *Rosa indica*) are also used in the extraction of essential oils [3]. In certain tropical countries, rose petals are used for producing rose water, "gulkand", rose petal jam etc., while rose hips are used for medicinal purposes.

4.1.1 Postharvest Quality Requirements

Factors such as cultivar, stem length, stage of maturity, flower size and conditions of the flower and the foliage, and the price of flowers influence consumer response in roses [6, 7, 9]. Laurie [8] listed the following quality components for rose:

1. Straight strong stem capable of holding the flower in upright position.
2. Uniform stem length-flowers with different stem length should not be mixed.
3. Size of flower representative of cultivar.
4. Uniform stage of development.
5. Flowers should be free from bruising injuries, petal discolouration and diseases and pests.
6. Good, healthy, normal foliage, the size representative of variety.

There are no uniform grades that are adopted universally. The purpose or market outlet, the country and the grower influence the grading procedure. Most growers in the USA grade their roses according to the stem length. Some producers use 5-cm increments from 25 cm onwards, while others employ 7.5- or 10-cm increments starting from 22.5 cm [8]. According to Prince and Robertson [2], only about 40% of the growers use mechanical grading techniques. This supports the need to establish industry-wide grades and standards for cut roses.

The European Economic Community (EEC) has introduced compulsory grades and standards in order to maintain uniformity in the grading procedure [1]. The EEC and SAF standards for fresh roses are given in Tables 4.1 and 4.5. It is important to note that none of the above grading systems considers the potential-lasting qualities of the cut flower [1, 12]. There are no research data available to correlate the stem length with the vase life of flowers. There

Table 4.1. The European Economic Community standards for fresh rose flowers [1]

Code	Stem length (cm)
0	5 or less
5	5 – 10
10	10 – 15
15	15 – 20
20	20 – 30
30	30 – 40
40	40 – 50
50	50 – 60
60	60 – 80
80	80 – 100
100	100 – 120
120	120 or more

Table 4.2. Utility index and relative importance of different purchasing components in rose [9]

Purchasing components		Utility index	Relative importance
Unit size	12	0.37	22.7
(No. of flowers/unit)	9	0.63	
	5	0.0	
Stem length (cm)	40	0.89	32.0
	56	0.34	
	66	0.0	
Price	Low	1.26	45.3
	Average	0.75	30.2
	High	0.0	
Stage of flower development	Tight bud	0.96	27.7
	Slightly open	0.64	64.0
	Fully open	0.0	
Cultivar	Junior Bridesmaid	0.09	45.1
	Sonia	1.75	
	Forever Yours	0.0	
Bent neck	None (5° bending)	0.58	17.4
	Slight (5 – 45° bending)	0.32	
	Extreme (45° bending)	0.0	
Petal discolouration	None (5%)	0.60	18.0
	Slight (5% – 20%)	0.34	
	Extreme (20%)	0.0	

is thus a need to develop subjective methods of grading in order to relate the vase life of cut flowers to the grading criteria.

Prince et al. [9, 10] evaluated several factors, like unit size, stem length, cultivar, condition of flower and price that control the consumer response in terms of their acceptability. The utility index (a relative utility score for purchasing components where lowest component value is set at a base value of 0) was used to measure consumer response. Cut roses sold in units of less than 12 flowers but more than 5 (particularly 9) were most acceptable to the consumer (Table 4.2). The price of the unit of flowers was found to be a major consideration in the purchase of smaller units. The stage of flower development at which cut roses had most acceptability was largely controlled by the cultivar. Sonia roses had highest consumer appeal when sold in an open bud stage, while Forever Yours roses were preferred in a tight bud stage. The extent of flower opening was found to be more important than other disorders like "bent neck". The sex of the consumer also seemed to influence the colour of the most accepted flower. However, in general, red Sweetheart roses had the greatest consumer acceptability, irrespective of sex.

4.1.2 Commercial Cultivars

Out of over 20000 cultivars that are currently on record, only few are under commercial cultivation. The choice of cultivar is influenced by consumer preference, the market outlets where they are being sold, growing conditions and keeping quality of cut flowers. However, cultivars such as Forever Yours, Samantha, Sonia, Baccara, Better Times, Junior Bridesmaid, Mary Devor, Belinda, Parada, Bridal Pink, Mercedes, Golden Wave, Visa, Cara Mia, Golden Times, Ilona and Carina are very popular in the rose trade.

4.1.3 Production and Distribution

Unlike in the cases of field crops, fruits and vegetables, there are no international organizations to compile and maintain the census on production and sale of different flower crops and ornamental plants.

The major producers of rose are the USA, Holland, Israel, the UK, Colombia and some of the South American and South African countries. Until very recently, Hybrid Teas, Floribundas and Sweetheart were grouped together when estimating the total area under cultivation, annual sale or production [13, 14]. Since these types differ in their growth habits, size, number of flowers produced per bush and net returns, figures arrived at by combining production under these types must be misleading.

Roses still maintain their first place in the USA, Canada, Holland, Israel and some of other European countries [1, 8]. While Holland, Colombia, Kenya and Israel are the leading exporters of roses in the world, the USA and some of the European countries are the major rose importers [12].

4.2 Physiology and Biochemistry

4.2.1 Growth and Development

Although rose is native to tropical countries, it can be easily grown in cooler climates. Light is the greatest limiting factor in the growth of rose plants in temperate countries where they are grown in greenhouses. Supplementary lighting has been shown to be beneficial in improving both productivity and quality of cut flowers during the winter in such climates [8]. Although day length apparently has very little influence on growth and flowering in roses, light intensity affects plant productivity. Cultural practices such as fertilizer scheduling, watering, cutting and pinching have to be planned based on the amount of light available during the growing period.

The growing temperature has profound influence on the quality of flower and also on the time of flowering. Although lower temperatures (below 13.2 °C) generally reduce plant productivity, they improve flower quality. In contrast, higher growing temperatures (above 18.2 °C) adversely affect the quality of the flower, but improve growth and flowering [15]. Similarly, day temperatures have also been found to have an impact on flower quality [15, 44]. Higher day temperatures (above 21.1 °C) reduce the mean stem length, flower fresh weight and weight-to-length ratio [15] (Table 4.3). According to Moe [44], the longevity was doubled when the growing temperature during the

Table 4.3. Effects of day and night temperatures on the yield and quality of Red Bird roses [15]

Temperature (°C) Day	Night	Total yield	Mean steam length (cm)	Mean fresh weight (g)	Weight/length ratio
At atmospheric CO_2					
21.1	15.5	992	26.1	24.8	0.95
26.7	15.5	1062	22.2	20.2	0.89
Supplemental CO_2 (1000 ppm)					
21.1	15.5	1280	29.7	27.7	0.93
26.7	15.5	1440	29.4	26.2	0.89
21.1	21.1	1570	25.4	20.6	0.81

last 3 weeks before flower cutting was maintained at 21° to 24 °C. However, low temperature reduction of longevity was eliminated by using preservative solution. In addition to the day and night temperatures of the aerial environment, root temperatures also influence the yielding ability and flower quality of greenhouse roses [16, 17]. A mean soil temperature of 18.3 °C was shown to produce a maximum number of quality flowers in Better Times roses [16].

Supplementary carbon dioxide (CO_2) up to 1000 ppm, when introduced into the greenhouse, improved the vegetative growth, yield and quality of flower [15, 18]. The prevailing temperature and light intensity play an important role in determining the beneficial effects of CO_2 enrichment [15, 18, 87].

4.2.2 *Senescence of Rose Cut Flowers*

Increase in respiration and hydrolysis of cell components are the two major metabolic changes that lead to petal senescence of cut roses [19, 20, 31]. Most of the enzymatic and other metabolic changes that occur during senescence have been shown to be related to the alteration in these two processes [20]. These changes enhance senescence in rose petals and decrease water uptake in cut stems by stimulating the formation of vascular occlusions in the stem [21–27]. These occlusions increase the resistance to water flow in the stem. Sterilization or treatment of stems with 8-hydroxy quinoline sulfate (8-HQS) was found to maintain xylem conductivity [27]. Microscope examination of stem sections from senescing cut roses revealed the occurrence of microbial occlusions (bacterial and fungal) and gum depositions [21, 23]. The microbial depositions were localized in the basal 2.5 cm of the stem, while gum deposition was dependent upon the depth of the holding solution [21]. Durkin and Kuc [26] hypothesized that vascular occlusions form at the basal end and progress acropetally with time. Acidic holding solutions (pH below 6.0) maintain conduction of water in cut stems by preventing the formation of vascular blockage [27].

Siegelman [20] demonstrated that increase in temperature of the ambient atmosphere stimulated respiration and senescence in Better Times roses. At

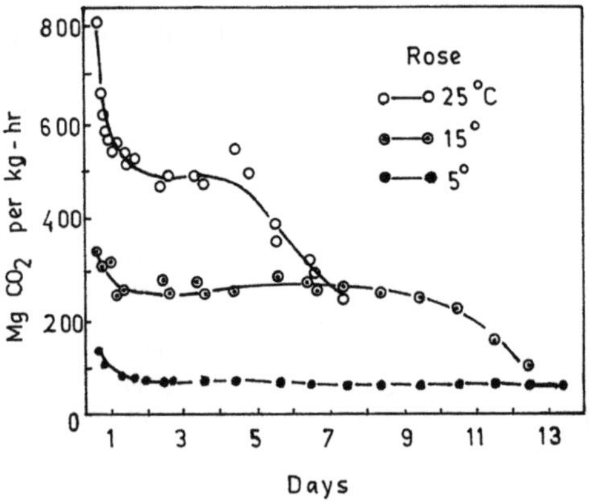

Fig. 4.1. Respiration of cut Better Times roses at different temperatures [19]

25 °C, the rose flowers showed a very high initial rate of respiration which slowed down rapidly for approximately 40 h before reaching a steady state. The respiration rate remained fairly steady for another 3 days, after which it began to decline. The flower appearance was satisfactory until the second decline (Fig. 4.1). In contrast, at 15 °C, the respiration rate reached a steady level within 15 h and remained constant for 8 days before the second decline began. The second drop in respiration rate was comparatively faster than the first one. There were no appreciable changes in the rate of respiration after an initial drop for about 15 h, and the flowers remained in excellent conditions throughout the period (13 days) of the study at 5 °C. Wulster et al. [28] illustrated the presence of cyanide-resistant respiration in Forever Yours rose petals harvested at the tight bud stage.

Although several investigations have been carried out to elucidate the senescence process in rose, they are still inconclusive [29–45, 84–86]. Detailed discussion on this aspect is given in Chapter 2.

Co-pigmentation of cyanidine, 3,5-glucoside with quercetin or kaempferol glycosides in petals of Better Times roses was found to depend upon pH. A slight change in pH of the vacuole influenced the co-pigmentation of anthocyanins. The blueing, a well-recognized phenomenon in senescing red roses, was demonstrated to be due to an increase in pH caused by breakdown of protein and accumulation of free ammonia. In certain varieties, the blackening or browning of petals was caused by the oxidation of flavones, leucoanthocyanins, phenols and accumulation of tannins [5].

4.3 Postharvest Losses

4.3.1 Extent of Losses

The extent of postharvest losses in roses varies according to the grower, the cultivar, and pre- and postharvest handling techniques. Cut flowers of rose are spoiled rapidly unless certain special treatments are given to protect them in the early hours after harvest. Although there are no comprehensive studies reported in the literature on the extent of postharvest losses in cut roses, it is generally accepted to be very high. Under normal situations, a loss of approximately 20% is very common, but it may be as high as 50% [1]. These estimates do not include crop that was never harvested from the field, or losses that occur at the wholesaler level. In certain tropical countries, where transport and marketing facilities are inadequate, losses due to wastage and reduction in quality may be substantially higher than the reported value, and in severe cases may be as high as 100%.

4.3.2 Nature and Causes of Losses

Pre-harvest growing conditions, harvesting techniques, and conditions during storage, pre-shipping and shipping period, and at the retailer and consumer levels influence the extent of postharvest losses.

Cultural Practices. The quantity as well as the quality of flowers produced in roses depends largely upon the pre-harvest growing conditions. Insufficient light levels and/or close spacings result in reduction of photosynthesis and affect the productivity of the bush. The flowers produced during winter months have comparatively shorter vase life than those produced during summer [1]. Inadequate carbohydrate supply has been shown to be associated with "bull head" and "blindness" in roses.

Growing temperatures above or below about 20 °C (night) reduce the potential-lasting quality of cut roses. Higher temperatures (21 to 24 °C) were found to decrease the degree of pigmentation and increase the blueing in petal tissues. Lower growing temperatures, on the other hand, increased the occurrence of "bent neck" in roses [44]. At the same time, if the buds before differentiation of floral organs were exposed to low temperatures (5 to 12 °C) they develop in to "bull heads", in which the flowers have shorter stalks and a larger number of petals and petaloids than normal flowers. The flower opening in "bull head" is not normal [43]. Low temperature exposure of flower buds also increases the activity of polymerase enzyme in petal tissue and cause accumulation of oxidation products of polyphenols. These polyphenols cause petal blackening [43]. The light intensity and temperature conditions during the last 2 to 3 weeks before harvest are most critical in determining postharvest quality. Low light intensity combined with high temperatures deplete the carbohydrates pool in the petal tissue and reduce the potential-lasting quality after harvest [8].

"Bull head" roses are also produced due to thrips infestation, insufficient carbohydrate supply to developing buds and hard pruning. Generally, these malformed flowers have shorter vase life [1].

The losses due to blind shoot formation vary with cultivars, age of the plant, growing conditions and time of the year. The amount of blind shoots under normal conditions varies from 35% to 40% at any time during the year.

Hard pruning techniques induce the formation of blind shoots in roses. Inadequate nutrition, unfavourable temperature and light conditions during flower bud initiation and early stages of development have been associated with blind shoot production in roses [8].

Scion-stock interactions also have considerable impact on the quality and quantity of flowers produced [46]. The size of the collar has been correlated with the quality of flower.

Several pesticides belonging to the organic phosphate group have been found to change petal colour in roses [8]. Red or pink flowers are most susceptible to organic phosphate pesticides, so that spraying these chemicals on the crop at the flowering stage may render the flowers unsaleable. Likewise, the use of phosphate aerosol in the closed greenhouse or irregular watering cause excessive loss of mature foliage, thereby indirectly affecting the flower quality. Most florist roses are susceptible to mercury vapours, which cause browning of petals and peduncle, and thereby render the flowers unsaleable. For this reason, mercury thermometers and other mercury-containing instruments should not be placed in rose ranges [8].

Bud harvesting may also increase the losses due to wastage and reduction in flower quality unless the grower fully understands the principle and takes adequate precautions to harvest the flowers at an appropriate stage.

In storage environments, high humidity (more than 97%) and low temperature are necessary in order to avoid losses in flower quality [1]. Temperature is a more important factor in storage of rose flowers than relative humidity. At high relative humidity, losses due to diseases are significantly higher unless proper care is taken.

Mechanical. Rose petals are very susceptible to bruising during and after harvest. Thorns on the stem very often tear off the outer petals when packaged. The stems in short-stemmed types have been shown to break easily in packages unless they are properly fastened together. Certain low-quality cellophane sleeves used to wrap the flower buds can also cause discolouration of petals in cut roses.

Physiological Disorders. "Bull head", frost or chilling injury, vascular plugging and nutrient deficiencies, especially of boron and zinc, are the most common physiological disorders associated with low flower quality in roses. Furthermore, sub-optimal growing conditions or extremely high light intensity and temperatures reduce the size of leaves, thereby decreasing the external appearance of flowers.

While extreme temperatures during shipping and/or storage impair the ability of flower buds to open normally and increase the occurrence of "bent neck", temperatures below 0 °C produce chilling injury. Figure 4.2 depicts changes in the temperature inside the box of cut roses during shipping. High temperature coupled with high relative humidity (standing water) induce microbial growth and cause damage to flowers and foliage. These postharvest diseases are hard to control after they have become established. Transportation

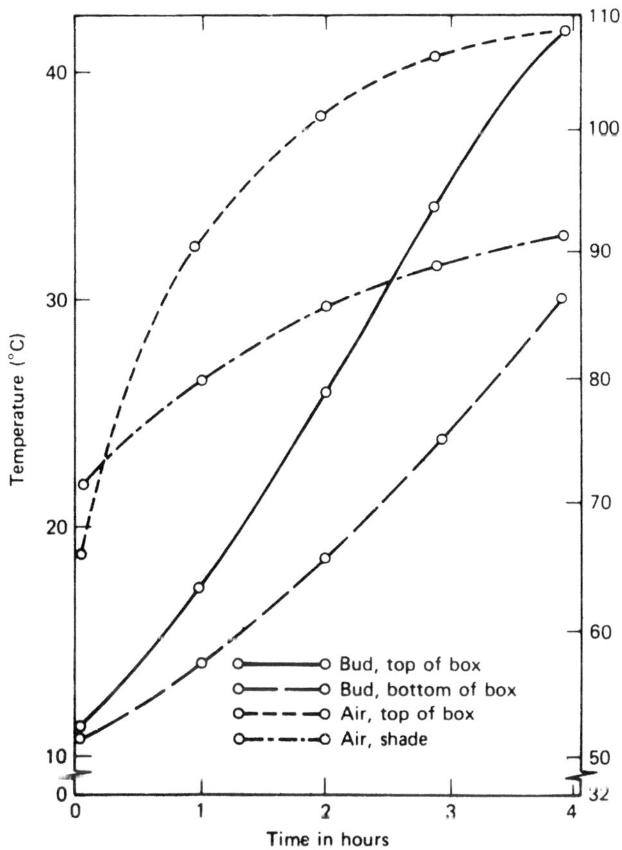

Fig. 4.2. Warming rates of Forever Yours rose buds in packages exposed to sunlight [84]

of cut roses with other products that produce ethylene will stimulate senescence processes in roses [1].

"Limp neck", another common disorder in roses, is caused by water stress in the area just below the flower bud. The affected flower buds bend downward due to the weight of the top. This is very common at the retail shop or at the consumer level. Insufficient food supply to flower buds has also been associated with this disorder [1].

Pests and Diseases. Powdery mildew is the most common disease affecting roses. It affects all parts of the plant, including the flower buds. The presence of whitish-coloured spores on the surface reduces the attractiness of the infected part and of the flower as a whole. Downy mildew and canker diseases in severe cases affect the productivity of the bush and the flower quality.

Black spot causes loss of foliage, and may pose serious problems unless it is controlled at an early stage. It affects stems, leaves and flowers. *Botrytis* fun-

Table 4.4. Relative quantities of ethylene produced by infected plant tissues [90]

Host tissue	Disease	Relative rate of ethylene production[a]
Normal tissue	–	1
Rose leaves	Black spot	5
	Botrytis	4
Carnation leaves	*Alternaria* leaf spot leaf spot	3
Rose leaves	Rust	2
Rose leaves	Powdery mildew	1

[a] 5 = maximum ethylene production

gus, on the other hand, attacks the flower and produces grey mould. In severe cases, it causes die-back of stems [88].

Verticillium wilt is not very common on roses. It affects mainly the leaves. The infected flowers or foliages produce two to five times more ethylene than normal tissues. The ethylene produced in infected tissues stimulates the senescent processes.

Virus infection causes chlorotic spots or streaks at or near the main veins. The infected leaves become distorted or dwarfed and reduce the overall appearance of the flower.

High temperatures and formation of water drops on the foliage and flowers during and/or after shipping encourages microbial growth in the packages. *Botrytis* rot is the most common postharvest disease in several flower crops and can grow even at −2 °C. High relative humidity and wounding of tissue increases the chances of *Botrytis* infection in storage environment. The pathogens, especially *Botrytis*, *Penicillium* and *Alternaria*, produce ethylene in greater quantities than is needed to accelerate senescence [89, 90] (Table 4.4). Wet conditions during harvesting increase the occurrence of *Botrytis* infection in storage. Microbial growth in the vase or on the stem enters the xylem and blocks the vessels, resulting in higher resistance to water uptake.

Insect Pests. Red spiders, aphids, and thrips are the important insect pests of roses. Thrips causes "bull head" formation in flowers, while aphids give an unsightly appearance to infected parts. Leaf rollers roll the petals and leaves which renders the flowers unsaleable. Other insects, like leaf-eating bees, cause serious damage to leaves and flowers. In addition to the physical damage to infected plant tissue, the wounds caused by insects stimulate ethylene production by the host, which in turn accelerates the natural senescent processes [89].

Root knot nematodes affect the quality of the flowers by reducing the vigour of plants. The infected plants suffer from malnutrition and inadequate water uptake.

4.4 Loss Reduction Biotechnology

4.4.1 Cultural Practices

Regulation of light and temperature conditions in the growing environment is beneficial in order to produce flowers having the best postharvest qualities. Supplementary lighting is necessary during winter months when solar radiation is inadequate to produce quality flowers. The optimum night temperatures for most of the florist rose cultivars is around 15.5 °C. The temperatures must not be suddenly increased or decreased, but rather changed gradually. Carbon dioxide enrichment coupled with proper temperature and light management is beneficial in obtaining higher yield of best quality rose flowers.

4.4.2 Harvesting and Packaging

From pruning to flowering, roses take about 38 to 42 days during spring or summer and up to 55 to 56 days in winter. They may require 3 to 7 additional days after soft pinch [8]. The stage of harvesting depends upon the variety and market distance. Generally, yellow varieties can be harvested in the tight bud stage, while rose or pink varieties do better when harvested later [8]. Flowers cut at too tight a stage fail to open under laboratory conditions. Most red and pink varieties are harvested when the sepals are perpendicular to the flower stem or the tip is pointing downward. Harvesting at early stages increases the incidence of "bent neck" in cut roses.

For local or short-distance markets the roses are cut when the outer petals start curling outward, while for distant markets they are harvested at an earlier (tight bud) stage. Precautions must be taken while harvesting. The shape of the bud in the axil of the leaf is used as a criterion for cutting rose flowers. The stems are cut to the leaf that has a rounded bud at the axil. The cutting practice varies with the grower, but during August to January it is desirable to maintain as much foliage as possible on the stem to prepare the plants for winter. From February onwards the flower stems can be cut "up and down" in several varieties [8]. Harvesting method will play a major role in the production of blind shoots in the coming season. Cutting the flower stems too low on the stem, especially in the cases of weak stems, will result in higher incidence of blind shoots in the following season [8].

The time of day when the flowers are harvested has direct bearing on the potential-lasting qualities. In roses, the flowers harvested at 16.30 h lasted 11% longer than those harvested at 8.00 h [1]. Flowers harvested in the late afternoon have a higher amount of stored carbohydrates due to that day's photosynthesis than those harvested in the morning. However, in most cases the mode of transport and availability of labour influence the time of harvesting.

The flowers harvested during May to September, with higher light intensities, have longer keeping qualities. Therefore, more flowers should be encouraged during this time of the year.

Table 4.5. The Society of American Florists' grades for cut roses [1]

Types	Grades minimum stem length (cm)		
	Blue	Red	Green
Hybrid Tea	56	36	25
Sweetheart	36	25	15

After harvest, the flowers are graded according to stem length and are grouped together in bunches of 10, 12 or 25 flowers. The Society of American Florists' (SAF) standards are the most common in the United States, while the EEC standards are compulsorily followed in Europe. Several growers in the USA have developed their own system of grading, which has largely replaced the SAF standards. The SAF grades for roses are given in Table 4.5.

Progressive growers in the USA and other developed countries use some kind of mechanized grading, but manual grading is still by far the most common method for grading of cut roses.

The graded flowers are then packed according to the market outlets. For long-distance markets, the flowers are packed in corrugated cardboard boxes. The flower bunches are wrapped in paper or cellophane sleeves and then placed in appropriate boxes. The paper pillows are placed at the bottom of the flower neck to avoid mechanical injury and the flowers are placed in layers in such a way that the successive layers rest on the stems of the flowers below.

The flowers must be packed well enough to avoid physical damage due to stacking or tossing of boxes during shipping. In warm weather, it is necessary to place shaved or crushed ice on the stem to prevent effects of desiccation or fluctuations in temperature [80]. In contrast, during the winter months, adequate insulation should be provided in the box to prevent freezing or chilling injury to flowers. The packaging materials must not absorb water from the flowers. The use of waxed cartons or polyethylene linning inside the box reduces the water absorption by the packaging materials and increases relative humidity inside the package. Wrapping of flowers with parchment paper or cellophane sleeve to prevent direct contact of flowers with the container protects the flowers from temperature fluctuations and desiccation during transportation or storage. Poor quality wrapping material can discolour the outer petals of flowers and reduce the marketability of cut roses [1, 8].

For local markets, roses are trucked in vases with water or preservative solution. Avoiding exposure of flowers to extreme temperatures is very important in maintaining flower quality.

4.4.3 Low Temperature Storage

Low temperature during handling, storage and marketing processes is an absolute necessity for roses [1]. The cut flowers are handled in two ways: (1) for local markets, and (2) for short-term storage (3–5 days). The roses are stored in preservative solutions at 0 to 1.6 °C. Immediately after harvest, the cut flowers are placed in preservative solution at 0.1 °C for at least 3 to 4 h prior to shipping. This removes all the field heat from the flowers and slows down the respiration process. Roses are then graded, packed and shipped or placed back into the cooler. For short-distance markets, flowers are shipped in preservative solution in refrigerated trucks.

Dry storage of flowers at 0 °C is beneficial for long-term holding. Cut roses can be stored for up to 15 days by this method. The flowers after harvest are placed in moisture-proof boxes and the field heat should be removed before sealing the containers. Placing the flowers in water before dry storage has adverse effects on post-storage vase life and increases the chance of blueing. At the end of long-term storage, the stems are recut before placing them in preservative solution for a minimum of 6 h to condition the flowers for shipping. When the flowers arrive at the retail shop, it is beneficial to harden them in preservative solution for 4 to 6 h at 0 to 1.6 °C [1].

4.4.4 Controlled Atmosphere (CA) Storage

Thornton [47] was the first to demonstrate the possibility of using controlled atmosphere (CA) for storing cut roses. His studies indicated that roses stored in CA conditions had better storageability and shelf life than those stored under low temperature conditions. Subsequently, other workers reported an increase in the postharvest life of cut flowers due to higher carbon dioxide concentration in the storage environment [48–51]. Thornton [47] was able to maintain the quality of cut flowers for 7 days (3 days in ambient air at 3 to 5 °C) and still obtain an additional 12 to 48 h of post-storage room temperature vase life (Table 4.6). CA-stored flowers opened rather more slowly than those stored in ambient environment at 3 to 5 °C. Neff [48] reported that carbon dioxide levels up to 5% to 25% and oxygen levels of 6% were necessary at the time of placing the flowers in the containers in order to effectively retard bud opening. Damage occurred to stigma, anthers and cambium tissues and necrosis of flower neck was noticed when they were held in 0.25% O_2. Best treatments were 0.5% O_2 and 5% CO_2 [49].

Carbon dioxide concentrations above 20 to 25% were injurious to certain rose varieties like Briarcliff and Mrs. F. R. Pierson, while Talisma flowers could

Table 4.6. Effects of controlled atmospheric storage on vase life of Better Times roses [47]

Storage conditions	Dry storage (−0.56 °C) Normal CO_2	3.3 °C 15% CO_2	3.3 °C stems in water
Optimum storage period days	15	7	3
Postharvest vase life at room temperature (days)	4	2.5–4.0	2
Total postharvest life (days)	19	9.5–11.0	5

be stored even at 30% without injury [47]. Toxic levels of CO_2 caused browning on the edges of the petals and failure of buds to open. These symptoms are accentuated at lower temperatures (0 °C). High concentration of CO_2 (above 15%) changes the levels of anthocyanin pigments in petals due to the reduction in cell pH [50]. The shift in colour from red or purple to green or blue has been associated with change in pH of the cell sap [91].

Recent investigations by several workers have indicated that low temperature dry storage extends the postharvest life of cut roses to several times that of CA storage and wet storage. According to Staby et al. [55], Forever Yours roses can be stored up to 4 weeks with 50% to 53% reduction in vase life at 0.5% to 8.0% oxygen. Tinga [50] compared CA storage (upto 15% CO_2) and dry storages in ambient air each at −0.56, 2.2, 4.4, and 10.0 °C. The oxygen levels in CA storage were reduced to 5%. CO_2 below 5% and O_2 above 15% did not improve the postharvest life of cut roses. In contrast, higher CO_2 with lower O_2 produced undesirable effects like petal and pedicel necrosis and discolouration. These investigations indicated that the best storage conditions were created by holding the flowers at ambient CO_2 and O_2 concentrations, at high relative humidity and in moisture proof containers. Hanan [51] also concluded that CA storage was not the best method of storage for cut roses.

4.4.5 *Modified Atmosphere (MA) Storage*

Small quantities of cut flowers can be sealed in a package of transparent films such as MSAT 300, and stored in that condition. This differs from CA storage in that the gas composition in the bag is not altered at the time of storage. With time, CO_2 accumulates inside the bag, while the relative oxygen concentration is depleted, with the result that the quality of the flowers is improved.

Roses sealed in MSAT 300 cellophane sleeves could be stored for 5 days at 4.5 to 7.2 °C without any loss in potential-lasting quality. In contrast, the flowers held in water in ambient atmosphere at 5 °C had the lowest post-storage vase life [52].

Although storage of flowers in 100% nitrogen has been possible in certain cut flower crops, it is not suitable for roses because of excessive blueing of petals. Furthermore, anaerobic respiration is commonly observed in nitrogen storage.

4.4.6 *Subatmospheric (Hypobaric) Storage*

Application of subatmospheric or hypobaric storage has been accepted by the fruit and vegetable industry, but its commercial utility in the floral industry is still questionable. It consists of storing flowers in cabins equiped to create; (a) low atmospheric pressure (101.3 kPa), (b) constant low temperature, (c) high relative humidity, and (d) a constant air change. As a system, hypobaric storage offers many advantages [55–59]. Bangerth [53] and Bredmose [54] demonstrated the possibility of storing cut roses under low pressure without severely affecting the flower qualities. Bangerth [53] succeeded in storing bud-stage

Table 4.7. Effects of low pressure (24 mm Hg) and normal storage of cut Merko Merceds and Sweet Promise Sonia roses on post-storage vase life (days) [54]

Treatments	Storage days			
	0	14	28	42
Merko Mercedes				
Normal pressure		9.0	5.6	5.5
Low pressure (LP)		14.4	9.6	7.3
Fresh control	15.8	15.4	15.2	13.1
Sweet Promise Sonia				
Normal pressure		8.7	3.5	4.3
Low pressure		10.7	6.5	6.1
Fresh control	12.2	11.3	11.7	12.1

Table 4.8. Effects of storage methods on the vase life (days) of rose cultivars [55]

Cultivar	No storage vase life	LP storage*		Normal	
		Vase life	%	Vase life	%
Forever Yours[a]	10.7	5.4	50	2.3	22
Sonia	7.2	3.6	50	2.2	30
Marinor	5.0	0.9	18	2.3	46
Samantha	11.9	5.8	49	3.2	26
Royalty[a]	13.1	6.7	51	3.2	24
Cora Mia	10.4	1.5	14	0.6	6
Town Crier[a]	3.9	1.1	29	0.3	9
Carina	6.8	2.5	37	2.9	43
Garnette	9.7	8.3	86	3.7	38
Belinda	8.0	4.9	61	2.5	31
Jack Frost	8.3	5.5	66	7.0	84
Spanish Sun[a]	7.9	7.8	99	4.9	62
Visa	11.5	6.7	99	4.9	62
Faberge	7.8	6.0	77	3.9	50
Golden Wave	7.6	3.1	41	2.6	34
Carte Blanche	6.0	2.5	42	3.5	58

* LP = low pressure
[a] Exhibited LP induced leaf damage percentage was calculated in relation to non-storage controls

cut Baccara roses at 4 °C and 75 mmHg for 49 days with subsequent vase life of 6 days in floral preservative at 25 °C. Burg [56] showed that bud-stage roses could be stored for 6 weeks without impairing their ability to open or subsequent vase life. Bredmose [57] was also able to store cut Belinda roses for 21 days and Merko Mercedes and Sweet Promise Sonia for 42 days with subsequent room temperature storage of 7, 7.3 and 6.1 days, respectively. Flowers stored under normal pressure had shorter lasting qualities than those from low pressure storage (LP) (Table 4.7). Wrapping the flowers during LP storage reduced the effectiveness of hypobaric conditions in Belinda roses. The flower size in this variety was improved by LP storage and by not wrapping the flowers with plastic.

Staby et al. [59] were able to store 16 cultivars of cut roses for 4–5 weeks with 61% of original or non-stored vase life (Table 4.8). He compared LP storage with normal refrigeration and low oxygen storage. Both pre- and post-storage treatments greatly influence any storage technology, including hypobaric storage, hence a total system approach must be utilized to achieve a success with any crop or storage method.

Several research workers [56–59] still question the feasibility of hypobaric storage for long-term holding of cut roses for the commercial trade. Staby et al. [55] reported foliar injury in Forever Yours, Town Crier, Royalty and Spanish Sun rose cultivars after 4–5 weeks of low pressure storage. The same cultivars from different growers also exhibited significant variation in the response to LP storage [55].

4.4.7 Use of Floral Preservatives and Chemicals

Use of chemical floral preservative solution is by far the most important means of improving the postharvest lasting quality of roses. Several chemicals have been tested for their bioefficacy in controlling microbial growth and vascular blockage and in extending the vase life of cut flowers in general and roses in particular [60–79]. Scholes and Boodley [65] evaluated more than 100 chemicals either alone or in combinations for extending the postharvest life of cut Velvet Times roses. They suggested a combination of 5% sucrose and 200 ppm of 8-hydroxy quinoline sulfate (8-HQS), and 50 ppm of silver acetate for extending the life of cut Velvet Times roses. Stem plugging in cut roses due to both microbial growth or deposition of gum was effectively controlled by placing the stems in preservative solution [69]. While the addition of 300 ppm of 8-hydroxy quinoline citrate (8-HQC) inhibited the growth of all fungi, bacteria and yeast commonly found in cut flower vase solutions, inclusion of 100 ppm sodium benzoate in a solution containing 8-HQC and sucrose prevented slime formation without having any phytotoxic effects [27, 69].

4.4.8 Control of Pests and Diseases

Apart from causing unsightly appearance, diseases and pests cause tremendous postharvest losses in quality and quantity of cut flowers. The affected plant tissues usually have a higher rate of ethylene synthesis, accelerating the senescent processes in roses. Postharvest diseases like *Botrytis, Penicillium* and *Alternaria* cause severe damage during storage. Therefore, integrated plant protection measures are necessary to produce and market quality cut flowers.

4.5 *Vase Life Evaluation in Roses*

In the past, the number of days a flower remains fresh or in acceptable condition was taken as a criterion for describing the keeping quality of flowers. These methods were used to evaluate the effectiveness of a given preservative in extending vase life [72, 81, 82]. Salinger [81] listed 11 criteria for evaluation of floral preservatives, including vase life, storage life, retention of flower colour, flower bud development etc. Other factors, like net water uptake in relation to vase life, have also been used by some researchers to determine the quality of flowers [82, 83]. All these methods provide subjective data for the evaluation of floral preservatives. Buys and Cours [83] developed an objective method in which water uptake was used to evaluate vase life of flowers. They established a mathematical relationship between cumulative water absorption of an individual flower and the time of the start of the experiment to give a model to estimate the vase life for an individual flower:

$$Y = AXB,$$

where A and B = constants to be estimated

Y = cumulative water absorption
X = time
B = specific parameter for each flower.

References

1. Staby GL, Robertson JL, Kiplinger DC, Connover CA (1978) Chain of life. Ohio State Florists Assoc Ohio State Univ, Columbus USA pp 71
2. Price TL, Robertson JL (1982) Marketing and postharvest management characteristics of the North American rose industry. HortScience 17(4):558
3. Bailey LH (1963) The standard encyclopedia of horticulture. McMillan New York USA
4. Gorer R (1970) Development of garden flowers. Eyres & Spottiswoode
5. Asen S, Norris KS, Steward RN (1971) Effect of pH and concentration of anthocyanin-flavanol copigment complex on the colour of Better Times roses. J Am Soc Hortic Sci 96(6):770
6. Parvin PE (1965) Studies on the effects of postharvest handling practices on keeping quality and marketability of Better Times roses. PhD Thesis, Michigan State Univ USA
7. Robertson JL, Chatfield LH (1981) Analysis of fresh flower merchandising. Ohio Agr Res Dev Centre Res Bull 1136:19
8. Laurie A, Kiplinger DC, Nelson KS (1968) Commercial flower forcing. McGraw-Hill Book, New York
9. Prince TL, Robertson JL, Chatfield LH (1980) Factors affecting the marketability of roses. J Am Soc Hort Sci 105:388
10. Prince TL, Robertson JL, Chatfield LH (1981) Factors affecting impulse purchasing of roses. Ohio Rep 66(1):11
11. Hutchinson N, Robertson JL (1978) An analysis of the demand for roses. HortScience 13:391
12. Staby GL, Robertson JL (1982) International movement of cut flowers. HortScience 17(5):729
13. Anon (1982) Floriculture crops production area and sales. Statistical Rep Serv USDA Washington DC
14. Anon (1970) Floriculture crops, production area, and sales. Statistical Rep Serv USDA Washington DC

15. Shaw RJ, Rogers MN (1964) Interactions between elevated carbon dioxide levels and greenhouse temperatures on the growth of roses, chrysanthemum, carnations, geraniums, snapdragons, and African violets. Flor Rev 135 (3486):23, 88; 135 (3487):21, 82; 135(3488):73, 95; 135(3489): 21, 59; 135(3491): 19, 37
16. Shanks J, Laurie A (1949) Rose root studies some of the effects of soil temperatures. Proc Am Soc Hortic Sci 54:495
17. Kohl HC, Fosler GM, Weinard FF (1949) The effect of several soil temperatures on flower production in roses. Proc Am Soc Hortic Sci 54:491
18. Lindstrom RS (1965) Carbon dioxide and its effect on the growth of greenhouse roses. Proc Am Soc Hortic Sci 87:521
19. Siegelman HW (1952) The respiration of rose and gardenia flowers. Proc Am Soc Hortic Sci 59:403
20. Siegelman HW, Chow T, Biale JB (1958) Respiration in developing rose petals. Plant Physiol 33:403
21. Lineberger RD (1974) Identification and characterization of vascular occlusions in cut roses. MS Thesis, Cornell Univ USA, pp 84
22. Mastalerz JW (1961) Life of carnations, roses, extended with preservatives, ionized water. Sci Farmer 8(3):10
23. Gilman KF, Stephonkus PL (1974) Vascular blockage in cut roses. J Am Soc Hortic Sci 97:662
24. Rasmussen HP, Carpenter JW (1974) Changes in the vascular morphology of cut rose stems: A scanning electron microscope study. J Am Soc Hortic Sci 97(5):454
25. Sacalis JN (1974) Inhibition of vascular blockage and extension of vase life of cut roses with an ion exchange column. HortScience 9:149
26. Durkin D, Kuc R (1966) Vascular blockage and senescence of cut rose flower. Proc Am Soc Hortic Sci 89:683
27. Marousku FJ (1971) Inhibition of vascular blockage and increased water retention in cut roses induced by pH 8-hydroxyquinoline citrate and sucrose. J Am Soc Hortic Sci 96:38
28. Wulster G, Sacalis J, Janes HW (1980) The presence of cyanide resistant respiration in petals of roses harvested at the tight bud stage. HortScience 15(3):315
29. Parups EV (1976) Acid and alkaline inorganic pyrophosphatase in senescing flowers of rose, carnation and chrysanthemum. Can J Plant Sci 56(3):525
30. Paulin A (1977) Metabolisme glucodique et proteique de la fleur d'acillet alimentee ou noa avec une solution de succhrose. Acta Hortic 71:241
31. Borochov A, Mayak S, Halevy AH (1976) Combined effects of abscicic acid and sucrose on growth and senescence of rose flowers. Physiol Plant 36:221
32. Mayak S, Halevy AH (1980) Flower senescence. In: Thimann KV (ed) Senescence in Plants. CRC Bocca Raton USA 131–156
33. Mayak S, Halevy AH (1974) The action of kinetin in improving the water balance and delaying senescence processes of cut rose flowers. Physiol Plant 32:330
34. Parups EV (1971) Disc electrophoresis of proteins of senescing and fresh leaves and petals of certain ornamental plants. J Am Soc Hortic Sci 96:168
35. Paulin A (1971) Influence de la composition de la solution nutritive sur la teneur en acides aminines libres et en ammoniac des petales de fleurs coupees. Ann Technol Agric (Paris) 20:283
36. Weinstein LH (1957) Senescence of roses. I. Chemical changes associated with senescence of cut Better Times roses. Contrib Boyce Thompson Inst 19:33
37. Schnabl H, Mayer I (1976) Dark fixation of CO_2 by flowers of cut roses. Planta 131:51
38. Halevy AH, Mayak S (1976) Senescence and postharvest physiology of cut flowers, part 1. Hortic Rev 1:204
39. Parups EV, Chan AD (1973) Extension of vase life of cut flowers by use of iso-ascorbate containing preservative solution. J Am Soc Hortic Sci 98:22
40. Borochov A, Halvey AH, Shinitzky M (1976a) Increase in microviscosity with ageing in protoplast plasmalemma of rose petals. Nature 263(5573):158
41. Borochov A, Halevy AH, Borochov H, Shinitzky M (1978) Microviscosity of rose petal's plasmalemma as affected by age and environmental factors. Plant Physiol 61:812
42. Sacher JA (1973) Senescence and postharvest physiology. Annu Rev Plant Physiol 24:197

43. Halevy AH, Zieslin A (1969) The development and causes of petal blackening and malformation in Baccara rose flowers. Acta Hortic 15:149
44. Moe R (1975) Effect of growing temperatures on keeping quality of roses. Acta Hortic 41:77
45. Staden OL, Molenar MH (1975) The effects of different mains water on the vase life of cut flowers. Vakbl Bloeisterij 30(42):21
46. Nikolova N (1974) A contribution to the question of scion-root stock combination in outdoor roses. Acta Pruhoniciana 32:21
47. Thornton NC (1930) The use of carbon dioxide for prolonging the life of cut flowers with special reference to roses. Am J Bot 17:614
48. Neff MS (1942) Effects of storage conditions on cut roses. Bot Gaz 103:794
49. Hurdenburg RE, Uota M, Parson CS (1967) Refrigeration and modified atmospheres for improving keeping quality of cut flowers. Proc 12th Int Congr Refrig 3:339
50. Tinga JH (1956) The effect of modified atmosphere storage at low temperature and treatments after low temperature storage which affect the keeping quality of cut flowers. PhD Diss, Cornell Univ, USA
51. Hanan JJ (1966) Controlled atmosphere storage of carnations. Colo Flower Grow Assoc Bull 262:1–5
52. Haugae AB, Bryant W, Laurie A (1947) Prepackaging of cut flowers. Proc Am Soc Hortic Sci 49:427
53. Bangerth F (1973) Zur Wirkung eines reduzierten Drucks auf Physiologie, Qualität und Lagerfähigkeit von Obst, Gemüse und Schnittblumen. Gartenbauwissenschaft 38:479
54. Bredmose N (1981) Effects of low pressure on storage life and subsequent keeping quality of cut roses. Acta Hortic 113:73
55. Staby GL, Cunningham MS, Holstead CL, Kelly JW, Kanjoyan PS, Eisenberg BA, Dressler B (1983) Storage of rose and carnation flowers. J Amer Soc Hortic Sci 109:193
56. Burg SP (1973) Hypobaric storage of cut flowers. HortScience 8(3):202
57. Bredmose N (1979) The influence of subatmospheric pressure on storage life and keeping quality of cut flowers of 'Belinda' roses. Acta Hortic Scandanavicat, 29:287
58. Carpenter JW, Dilley DR (1975) Investigations to extend cut flower longevity. Michigan State Univ Res Rep 263:1
59. Staby GL, Cunningham MS, Eisenberg BA, Bridgen MP, Kalley JW (1979) Low pressure and controlled atmosphere storage of carnation and roses. HortScience 14(3):446
60. Coorts GD, Gartner JB (1963) The effects of various solutions on keeping quality of Better Times roses with or without hooks. Proc Am Soc Hortic Sci 83:833
61. Scholes JF (1963) Some of the effects of various chemicals on the postharvest physiology of Valvet Times roses. MS Thesis, Cornell Univ USA
62. Aarts JFT (1957) Over de houdbaarheid van snijbloemen. Meded Landbouwhogesch Wageningen 57:1
63. Weinstein LH, Laurencot HJ Jr (1963) Studies on the preservation of cut flowers. Contrib Boyce Thompson Inst 22(2):81
64. Ryan WL (1957) Flower preservatives using silver and zinc as disinfectants. Flor Rev 121 (3129):59
65. Scholes J, Boodley JW (1964) Improved lasting life of Velvet Times roses with chemicals. NY State Flower Grow Assoc Bull 224:1
66. De Stigter HCM (1981) Ethephon effects in cut sonia roses after pretreatment with silver thiosulfate. Acta Hortic 111:27
67. Kofranek AM, Kohl HC, Kabotai J (1975) Stabilized chlorine compound as a vase-water additive. Calif Flower Nurs Rep Jan-Feb, 5–7
68. Kofranek AM, Kohl HC, Kabota J (1974) A slow release chlorine compound as vase-water additive. Flor Rev 154(4000)21:63
69. Marousky FJ (1976) Control of bacteria in cut flower vase water. Proc Fla Sta Hortic Soc 90:234
70. De Stigter HCM (1981) Water balance aspects of cut and intact Sonia rose plants and effects of glucose, 8-hydroxyquninoline sulfate and aluminium sulfate. Acta Hortic 113:97
71. Mayak S, Halevy AH (1974) The action of kinetin in improving the water balance and delaying senescence process of cut rose flower. Physiol Plant 32:330

72. Beever DJ (1980) Use of plant food in the postharvest storage of carnation. Acta Hortic 113:130
73. Metzger B (1972) The effects of growth regulators, 8-HQC and sugar on cut rose vase life. Colo Flower Grow Assoc Bull 266:1
74. Coorts GD, Gartner JB, McCollum JP (1965) The effects of senescence and preservatives on respiration of cut flowers of *Rosa hybrida* 'Valvet Times'. Proc Am Soc Hortic Sci 86:799
75. Venkatarayappa T, Murr DP, Tsujita M (1981) Effects of cobalt and sucrose on the physiology of cut Samantha roses. HortScience 56(1):21
76. Asen A, Davidson OW (1952) Boron requirements of greenhouse roses. Proc Am Soc Hortic Sci 60:439
77. Yehosua SB, Juven B, Fruchter M, Halevy AH (1966) Effects of ethylene oxide on opening and longevity of cut rose flowers. Proc Am Soc Hortic Sci 89:677
78. Coorts GD, McCollum JP, Gartner JB (1965) Effect of senescence and preservative on mitochondria in flowers of *Rosa hybrida* Velvet Times. Proc Am Soc Hortic Sci 86:797
79. Conrado LL, Shanhan R, Eisinger W (1980) Effects of pH, osmolality and oxygen on solution uptake in cut rose flowers. J Am Soc Hortic Sci 105(5):680
80. Mastalerz J, Beever DJ (1963) Flowers on ice – how to do it right. In: MN Rogers (ed). Living flowers that last. Univ Mo 58
81. Salinger JP (1975) Criteria for the evaluation of postharvest senescence of cut flowers. Acta Hortic 41:207
82. Systema W (1975) Conditions for measuring vase life of cut flowers. Acta Hortic 41:217
83. Buys CA, Cours HG (1981) Water uptake as a criterion for the vase life of cut flowers. Acta Hortic 113:127
84. Maxie EC, Farnham DS, Mitchell FG, Sommer NF (1974) Temperature management effects on quality of carnation flowers and rose buds Calif. Agric 28(1):6
85. Kaltaler REL, Steponkus PL (1976) Factors affecting respiration in cut roses. J Am Soc Hortic Sci 101(4):352
86. Halevy AH, Mayak S (1975) Interrelationship of several phytohormones in regulation of rose petal senescence. Acta Hortic 41:103
87. Matson RH, Widmer RE (1971) Effects of carbon dioxide during growth on vase life of greenhouse roses (*Rosa hybrida*). J Am Soc Hortic Sci 96:284
88. Frankin RA, Huges IK (1982) Rose growing in Queensland. Queensland Agric J 108(4):245
89. Williamson CE (1950) Ethylene, a metabolic product of diseased or injured plants. Phytopathology 40:205
90. Williamson CE (1963) Plant diseases that affect flower quality. In: Rogers MN (ed) Living flowers that last. Univ Missouri Columbia USA
91. Thornton NC (1934) Carbon dioxide storage VII changes in flower colour as an evidence of effectiveness of carbon dioxide reducing the acidity of plant tissues. Contrib Boyce Thompson Inst 6:403

Chapter 5 Chrysanthemum

5.1 Introduction

Florist chrysanthemum is grown throughout the world both for cut flowers and as potted plants. This plant has been studied intensively by scientists and growers, with the result that the crop can now be grown to precise schedule at any time by controlling the growing to precise schedule at any time by controlling the growing conditions. The ability of the producer to obtain a quality crop of chrysanthemum depends upon understanding the complex interaction between the plant and its environment.

In the USA, chrysanthemums are produced for daily markets, but there is no heavy demand for any particular holiday [3]. However, in Europe, the demand for chrysanthemum is greatest for All Saint's Day. As a cut flower, it still maintains first place in Germany, and second place in Holland and Italy [1, 2].

The classification system most popular among growers and customers is based on the habit of growth or method by which they are handled. In this method all the florist chrysanthemum cultivars are grouped into standard, spray and disbudded types. Standard are the large-flowered cultivars with one terminal flower produced per stem. The lateral buds and side shoots are disbudded to encourage the growth of the central bud. Mainly incurved and reflexed types are used as standards, while spiders and fujy cultivars are also grown as standards. Of these types, incurved standards are produced in greatest quantity. The flower size in the standard group varies from 10 to 15 cm or more. If the size of flower of these types is smaller than 10 cm, they are produced as disbuds. Spray types, on the other hand, are smaller-flowered cultivars in which spray of flowers per stem is encouraged by pinching the terminal flower bud. The pompoms are the most common spray chrysanthemums produced in the USA and Europe. Other groups grown as sprays include anemones, singles and decorative. Spray chrysanthemums are sold on a weight basis. All these types are produced all the year round in Florida and California, where favourable growing conditions exist [3].

5.1.1 Postharvest Quality Requirements

Cut chrysanthemums are sold according to the size of the flower or weight of flowers and stems. The standard chrysanthemums which are generally used for bouquets are sold singly, based on the size of the flower and stem length [4].

Table 5.1. The SAF* grades for standard chrysanthemums [3]

Quality parameters	Grades			
	Blue	Red	Green	Yellow
Minimum stem length (cm)	75	75	60	60
Minimum flower diam (cm)	15.0	12.5	10.0	–
Stem strength	Strong	Strong	Strong	–

* Society of American Florists

Table 5.2. Polish grades for cut chrysanthemum [5]

Parameters	Grades	
	I	II
Flower diam (cm)	13	8
Stem length (cm)	60	45
Permissible faults	–	Slight stem deformation

Table 5.3. Metric grade specification for spray chrysanthemum [2]

Grade	Stems/sleeve	Specifications
Gold	10	6 or more flowers out and some to come
Silver	15	4–5 flowers out and some to come
Bronze	20	3 flowers out and some to come
Make ups	–	All stems not covered above, filling sleeves to the same extent as in other grades

Table 5.4. SAF inspection standards for cut chrysanthemums [3]

1. Bright, clean, and not badly wilted foliage and flowers
2. Flowers with similar varietal characteristics bunched together
3. Shape and size of flower representative of the cultivar
4. Fairly tight flower with undeveloped centre
5. Flowers and foliage free from only defect, injured, dirt or foreign materials
6. Free from discolouration, nutritional, chemical or mechanical abnormalities
7. Fairly straight, stiff stems capable of supporting the flower in an upright position
8. Foliage stripped from not more than the lower one third of the stem
9. Stem length according to the requirement of the grade

There are no compulsory grades to be followed by the growers in the USA. However, most growers grade their standard mums according to the standards suggested by the Society of American Florists [3] (Table 5.1). The Polish grades for standard mums is given in Table 5.2. In Britain, standard mums are graded according to flower diameter [2].

The spray mums in the USA are marketed on a weight basis (usually 9 or 16 oz), while in Britain there are established grades [2] (Table 5.3).

The Society of American Florists has established certain inspection standards to help growers and consumers to evaluate the quality of the product (Table 5.4).

The length of the peduncle is an important postharvest quality consideration in spray types. A well-formed flower placed at right angles to the stem and with adequate stalk length is always preferred in the market. Some of the pompoms, due to their brittle peduncle, crack or break just below the flower head, and may add to reduction both in quality and quantity of saleable flowers [4].

5.1.2 Commercial Cultivars

Standard Chrysanthemums. In both the USA and European countries, there are very few standard cultivars that are suitable for year-round production because the varieties from 12-, 13- and 14-week response groups are not adaptable for this purpose [2]. The varieties for all-year production must have strong vegetative growth, which responds to day length very quickly without affecting the flower quality. Secondly, for this purpose the flowering should spread over a fairly short period so that once-over harvest is feasible. The 8- to 10-week response groups offer these advantages for year-round production.

Indianapolis White and its sports are very popular in the USA. These cultivars need interrupted lighting during winter [3]. In Europe, Indianapolis varieties have been found to be the best for October-November and April-May flowering, while they need supplemental lighting from December to March. The petals are damaged due to scorching if the summers are too bright. For winter, under European conditions, cultivars like May Fred Shoesmith and their sports are more suitable than others [2].

In Britain, the cultivar Snowdon is the most common in year-round production. During summer, slight crinkling of the petals is observed in this variety. Other cultivars, like Escapade, Princess Anne, and Woking Scarlet, are also grown to some extent. Most cultivars of American origin have more dependable vegetative characteristics, while British mums have vigour, wider range of colours and forms. The German consumers prefer larger-flowered varieties than is usual otherwise in the USA and Europe.

Spray Types. The choice of cultivar depends upon the market requirement, due to which there is lot of variation in the varieties grown in different countries. For example, the mums produced in Holland are sold in Germany. This necessitates the production of varieties that suit both the domestic as well as foreign markets. In Holland and Germany the spray varieties from the spider group are more popular, while in Denmark, Sweden and Finland, single varieties are in greater demand [2]. In contrast, decorative spray types are particularly favoured in Britain.

An ideal spray variety should have 6 to 12 equal-sized lateral flowers at a similar stage of development and arranged attractively on the stem. In the

Table 5.5. Postharvest qualities of cut chrysanthemum cultivars [5]

Cultivar	Stem length (cm)	Flower diam. (cm)	Foliage
Alec Bedsar Gold	75	13	
Bornholm (and sports)	80	11	
Clingendaal (and sports)	70	14	Fragile
Diplomat (and sports)	70	14	
Escort (and sports)	100	12	Soft
Hartman's Perfection	70	16	
Jacob's Lane	85	15	Fragile
Maureen Plummer	80	15	Soft
Pandion	85	15	
Pink Pride	70	11	Fragile
Regalia Purpeen	75	14	

USA, spray varieties are sold on a weight basis, and a bunch of 9 oz should contain at least four to five good flower stems [3]. Mostly the cultivars belonging to 9- or 10-week response group are suitable for year-round production. Again, different varieties are preferred during different seasons of the year. For example, most of the autumn varieties are not liked by the public during spring, when only few colours are required. Likewise the varieties for winter should be less sensitive to light and should be able to set buds under cool conditions. In contrast, the varieties for summer should be able to grow uniformly under dry, warm weather. Most of the florist chrysanthemums vary in their postharvest qualities [5] (Table 5.5).

5.1.3 Production

Chrysanthemum as cut flower maintains first place in Germany, and second place in Italy, Holland, the USA and Britain [1].

5.2 Physiology and Biochemistry

5.2.1 Growth and Development

The intensity and duration of light has a profound influence on the quantity and quality of flowers produced. Most chrysanthemum cultivars tend to grow more rapidly at higher light intensities, especially under optimum growing temperatures. The photosynthetic rates increase with radiant energy, reaching a maximum around 3000 to 10000 ft. c. [3]. The response of chrysanthemums to light depends largely upon the cultivar, temperature, carbon dioxide concentration, water and mineral supply. Generally, under favourable growing temperatures, the petal carbohydrate content also increases with radiant flux intensity, which has been shown to have direct correlation with postharvest lasting qualities in flowers. The carbohydrate content in petals is reduced at high temperatures combined with low light intensities. Therefore, the last 2 weeks in the greenhouse before harvesting are critical in determining the subsequent flower quality. Mastalerz [6] demonstrated that lowering the light intensity to one half the optimum levels during the last 2 weeks in the greenhouse significantly reduced the vase life of cut chrysanthemum after harvest. It also adversely affected the storageability of flowers. Seasonal variations in light intensity are also reflected in the flower quality after harvest. In general, flowers produced during spring and summer months have longer shelf life than those produced during winter months [4]. The quality of foliage and flower stems is also affected by low light intensities [7].

Flowering in most of the florist chrysanthemum cultivars is initiated under short days [9–16, 27–29]. These cultivars are classified according to the number of short days required for flowering, into 6- to 16-week response groups. The cultivars belonging to 9-, 10- and 11-week response groups are particularly suitable for year-round production, whereas 6-, 7- or 8-week cultivars have been found to be suitable for growing as garden mums [2, 3]. The differences in number of short days required for flowering of various cultivars are related to the rate of flower development [8]. These varieties differ in their critical day-length requirements for flower initiation. A 6-week cultivar needs a longer day length than the late-flowering types [8]. The flower bud development is more rapid at higher temperature, irrespective of day length [9–12].

Although the physiological transformation of the vegetative bud into flower bud takes place very early in the induction period (4–5 short day treatment), complete flower initiation and the development to the visible bud stage take nearly 21 short days. The photoperiod plays a critical role during this period of flower bud development. The exposure of the bud to longer day lengths at this stage produces crown bud (a flower bud in which floret primordia are not initiated). The lateral buds grow into vegetative shoots and the overall quality of the flower is reduced if the long days are continued beyond the critical limit.

Interrupted lighting, in which short days are followed by a few long days and then returned to short days until harvest, is utilized to improve the size and colour of flower and the number of ray florets per flower in standard types [13, 14]. It also regulates the arrangement of spray of pompom types. Post and Lacy [15] demonstrated that in the case of standard mums, providing 12 short days followed by 10 long days and then short days improved the flower quality by increasing the number of ray florets per bloom. These investigators further showed an improvement in the quality of sprays by subjecting the plants to 12 short days followed by 10 long days (or 4 to 5 short days, followed by 15 to 20 long days), and then returning to short days.

The diameter of standard mums can also be improved by after-lighting, in which the plants are lighted continuously starting from 4 weeks of short day treatment until harvest, the flower weight or the number of ray florets is not affected by after-lighting treatment [13] (Table 5.6). It is important to note that both these lighting treatments delay anthesis.

Although flowering in chrysanthemum is governed by photoperiod, many cultivars tend to deviate in their critical day length requirements at different

Table 5.6. Effects of after-lighting and interrupted lighting on flower quality in standard mums [13]

Cultivar	Weeks of short days	Delay in maturity	Flower diameter (cm)	Fresh wt. (g)	Longest floret (cm)	Number of ray florets
White Pink	4	6	10.5	17.4	6.75	328
Chef	5	2	9.75	18.0	6.75	315
	Cont. S.D.	0	9.50	17.9	6.50	320
	IL	8	9.25	20.0	6.25	393
Giant 4	4	16	10.75	18.0	8.75	303
	5	15	10.75	16.8	8.00	295
	Cont. S.D.	0	10.00	13.2	6.75	292
	IL	17	10.50	22.6	7.50	463
Mefo	4	13	11.5	20.0	7.65	373
	5	12	12.00	22.8	7.65	378
	Cont. S.D.	0	11.00	20.2	6.75	368
	IL	8	11.00	38.2	7.5	635
Yellow	4	7	13.50	31.0	8.50	343
Shoesmith	5	5	13.25	22.1	7.50	359
	Cont. S.D.	0	11.25	22.7	7.50	350
	IL	8	11.00	25.2	7.50	516

Cont. S.D. = Continuous short days
IL = Interrupted lighting

temperatures [9–12]. The flowering time is also shifted from that of normal ones, depending upon the fluctuation in temperatures [16]. The optimum temperature for chrysanthemum is 15.6 °C [12]. Any deviation from optimum temperatures results in delay in flower initiation and/or development and affects the size and form of flowers. Several varieties, especially from the Indianapolis series, produce a higher percentage of quilled florets at lower temperatures [17]. These malformed (quilled) florets impart a flatter appearance to the flower head, and as such are not preferred by the consumer. Miller and Kiplinger [17] suggested that low temperature was more critical than photoperiod in producing this disorder.

Several studies on root zone warming (RZW) have indicated its possible use in improving the yield as well as the flower quality in mums [20–24].

Carbon dioxide enrichment usually increases stem length, fresh weight of stems and flowers and total yield. However, the response of plants to CO_2 enrichment depends largely upon the cultivar and age at the time of treatment [18, 9, 28, 29].

5.2.2 Senescence of Cut Chrysanthemum

The published literature on different aspects of flower senescence in cut chrysanthemum is scanty and, as such, it is difficult to document all the biochemical and physiological changes in ageing chrysanthemum inflorescence. Basically, the processes leading to irreversible wilting and senescence of cut mums are similar to those in other flowers.

5.3 Postharvest Losses

5.3.1 Nature of Losses

Postharvest losses in cut mums result to a large extent from improper handling of blooms after harvest or during transportation. Due to their fragile nature, the foliage and florets in chrysanthemum are vulnerable to bruising injuries. These injuries enhance ethylene synthesis in the flower tissues and cause premature wilting of inflorescence. The removal of the stems from a permanent supply of carbohydrates and water causes rapid wilting of foliage and flowers. If the wilting is prolonged beyond a certain limit, the flowers fail to regain turgour when returned to water; this also adds to the postharvest shrinkage.

5.3.2 Causes of Losses

Cultural factors (light, temperature, pinching, water stress), mechanical damages during and/or after harvest, physiological disorders, diseases and pests are mainly responsible for the enormous postharvest shrinkage that exists in cut mums. The extent of losses due to mechanical damage (broken and shattered blooms, torn out or crushed leaves etc.) depends largely upon the method of packaging and transportation [30, 31]. Krahn [30] estimated the losses due to mechanical damage in the form of shattered blooms alone to be around 15%. If we take into consideration all injuries and damage that occur during postharvest handling, then these losses will be much higher. Uncooled flowers are more vulnerable to mechanical damage than precooled ones. The mode of transportation has a direct bearing on the mechanical injuries. In general, surface transportation causes more damage to cut flowers than air shipment. Stem cracking just below the flower head, common in certain cultivars, also adds to these losses.

Apart from mineral deficiencies, premature budding, sun-scald, quilling of florets, crown bud formation and heat delay in flowering are some of the common physiological disorders that are encountered in cut chrysanthemums [25–27, 32–37]. These disorders enhance the losses due to wastage of flowers or decrease in the quality of the finished product.

Formation of flower buds during long day treatment or as soon as the short day treatment began is commonly referred to as premature budding [25–27]. These buds fail to develop into quality flowers. The initiation of premature buds depends upon the physiological age of the plant, and environmental factors determine the timing of bud initiation [32].

The extent of stem lignification to a great extent determines the ability of the cut stems to conduct and transport water in chrysanthemums, roses and carnations. Older stem tissues at the lower end of the flower stalk reduce the water uptake by the flower. Therefore, cutting the flower too low impairs the ability of stems to conduct water and hampers the effectiveness of preservative solutions in preventing wilting of flowers [59]. Blockage of vascular tissues by metabolic byproducts or microbial growth also reduces water absorption in the stem and causes premature wilting of flowers and foliage.

Boron deficiency is commonly reflected in severe petal damage [65, 67–70]. The boron levels that are considered to be adequate in many floral crops cause severe petal injury due to deficiency, but become extremely toxic to foliage and flowers in quantities just above the optimum [68, 71].

Chrysanthemum flowers, due to their delicate nature with large total leaf area and succulent foliage, are prone to wilting caused by imbalances between transpiration and water uptake. In severe cases, wilting can result in irreversible damage to cut flowers in terms of reduction in storageability and flower quality. Furthermore, foliar chlorosis is frequently observed during long-term storage, especially when flowers are held in the dark or at high temperature [66].

Losses due to pests and diseases can be very high. Apart from physical injury to the plant tissues, attack of plants by insects and diseases stimulates ethylene synthesis in the host tissue and can cause senescence of flowers.

Insects like aphids, leafminers, cutworms, cyclamen mite, grasshoppers, leaf rollers, red spiders, thrips, and nematodes attack chrysanthemum plants [42, 43]. These insects disfigure the young growth or chew the foliage or flower tissues. Leafminers, which make tunnels between the upper and lower epidermis, are very difficult to control, as the damage is done before the insect is noticed.

Among diseases, bacterial stem rot, *Botrytis* blight, white rust, *Verticillium* wilt, *Pythium* rot (stem and root) and *Ascochyta* (ray) blight are very common in chrysanthemum [38–40]. Certain cultivars of florist chrysanthemums are also susceptible to stunt and yellow virus [40, 41]. The intensity of disease infection depends upon prevailing environmental conditions like temperature, relative humidity and soil moisture content. *Botrytis* causes the petals to become brown and water-soaked [38], and also attacks the flowers and foliage during storage. White rust caused by *Puccinia horiana* is a very serious disease in Europe [45]. It attacks stems, leaves, bracts and ray florets and reduces yield and flower quality [44, 45]. Certain varieties are more susceptible to this disease than others. Apart from direct damage to the infected plant, these fungi produce two to five times more ethylene than normal tissues, causing premature senescence.

5.4 Loss Reduction Biotechnology

5.4.1 Cultural Practices

The quality of chrysanthemum flowers can be improved by manipulating photoperiod and temperature, carbon dioxide enrichment, use of growth regulators, and water management techniques [13–15, 46–58]. An integrated approach utilizing all of these ways is required to produce quality crop.

5.4.2 Harvesting, Grading and Packaging

The stage of flower development at the time of harvesting determines success with bud opening and subsequent postharvest vase life. According to the old method of harvesting, the flowers were cut when the inflorescence was fully open, but the flowers are presently harvested when they are still in the green bud stage. Cultivars from the single group are harvested before the pollens from the outer row of disc florets are shed. Anemones should be cut before the central cushion in the topmost flower is fully developed, while the decorative types perform better when they are harvested after full development of the central florets of the topmost flower. In contrast, standard mums are harvested when the outer ray florets cease to elongate and the centre of the flower is well developed [3]. The stage of harvesting depends upon the market outlet where the flowers are being sold and the facilities available to the grower [4].

Once-over harvest or one-time harvest is commonly followed in cut mum production. The advantages of once-over harvest are well established. When combined with bud harvesting, it gives a grower more flexibility in marketing his crop [59–64]. With this technique, the crop can be saved from adverse environmental conditions like frost, rain or low light by being harvested at an early stage. *Botrytis damage* can be minimized by cutting the crop in bud stages and letting it open in preservative solution (8-HQC+sugar) under laboratory conditions. The losses due to bruising and floret shattering during harvesting and postharvest handling can be completely eliminated by this method of harvesting. Since mechanical harvesting is possible with this technique, the cost and time involved are negligible as compared to conventional harvesting. However, once-over harvest requires uniform flowering, selection of suitable cultivar, proper spacing and grading of cuttings at the time of planting for production of uniform-sized flowers at the time of harvest [59, 64].

The condition of the stems at the time of harvest was found to have an influence on the rehydration of cut chrysanthemums [59]. Marousky [62] demonstrated a direct correlation between the softness of the stem tissue at base at the time of harvesting and changes in flower fresh weight. The ability of cut stems to rehydrate has tremendous effect on the postharvest lasting quality of the cut flower.

The flowers are graded according to the standards established by the SAF (Table 5.1) or EEC (Table 3.1). The foliage from the lower third of the stem is removed to avoid its rotting in holding solution. The method of packaging depends upon the type of market outlets being served. In the past, low density display boxes were used for packing chrysanthemums [3]. These boxes are expensive to make and transport, and involve a lot of time in packing [2]. Further, they do not provide adequate protection to the flowers. In recent years, cardboard boxes have been used for packing cut mums. The size of the boxes varies with the country and the type of mums. The spray varieties are packed in boxes measuring $80 \times 50 \times 30$ cm, while $91 \times 43 \times 15$ cm boxes are used for standard types [2]. The boxes are lined with polyethylene film to maintain higher humidity inside the box. This will also protect the flowers from coming into contact with the container.

In the case of standard mums, the flowers are packed separately by covering them with protective sleeves (or wax paper) or sealing them in plastic bags. Some growers also use plastic mesh sleeves to cover the blooms. The plastic bags or mesh sleeves were found to be most effective in protecting the blooms from mechanical damage. Other wrapping materials like polysterol, polypropylene and perforated polypropylene sheets (700 holes/m^2) have also been used to cover the blooms in boxes. Perforated polysterol sheets are equally as effective as plastic bags or mesh sleeves as wrapping material for cut mums.

The blooms are placed in boxes according to grades. For long-distance sales outlets, the boxes will have to be protected against extreme temperatures. The use of pillow or "cushions" below the flower heads in the box will prevent crushing or shattering of florets or breaking of flower stems. For local markets the flowers can be trucked in vases with preservative solution or water. Cultivars belonging to spray types are usually packed in bunches in the same way as standards.

5.4.3 Low Temperature Storage

Precooling before shipping the flowers is an absolute necessity in cut mums. Wiersama and Boyer [72] demonstrated that uncooled flowers packed in normal florist packages are exposed to higher temperatures during storage. The temperature of uncooled flowers in packages rose to 35 °C during an unrefrigerated transport, while the increase was only 10 °C in the case of precooled flowers. The rate of heat removal from uncooled flowers during refrigerated transport was determined to be 1 °C per 12 h, indicating that refrigerated transport cannot by itself compensate for precooling. Ethylene has been found to be many times more active at 35° than 10 °C [75]. The room temperature shelf life was extended by shipping the pre-cooled flowers in refrigerated trucks [75].

The optimum temperature for long-term holding of cut mums is 2 to 5 °C. Generally, storage temperatures below 1 °C cause freezing injury. Maintaining high relative humidity (97% to 98%) is beneficial for cut mums [3]. Under these conditions, mums can be stored for 40 days without any appreciable loss in quality or shelf life.

5.4.4 Controlled Atmospheric (CA) Storage

Although the benefits of controlled atmosphere in extending the postharvest shelf life of flowers were demonstrated by several workers [76–79] in the 1930s and 1940s, it is not commercially employed for holding chrysanthemum flowers. Tinga [79] stored chrysanthemums at different temperatures (−0.56, 2.2, 4.4, and 10 °C) in ambient air or in elevated carbon dioxide concentrations up to 15% and reduced oxygen levels to 5%. The results indicated that slight changes in the gaseous composition of the storage atmosphere did not extend the shelf life of cut mums, but major changes (higher levels of CO_2 and lower levels of O_2) resulted in undesirable effects like changes in petal colour or petal and pedicel necrosis [79]. These results clearly indicate that CA storage is not useful for long-term storage of flowers.

5.4.5 *Modified Atmospheric (MA) Storage*

The levels of carbon dioxide in the storage atmosphere can also be altered by utilizing modified atmospheric storage, in which the flowers are packaged in special quality plastic bags and stored [79, 80]. With this method, pompom chrysanthemums could be stored for up to 7 days without any loss in poststorage vase life at room temperature. However, MA storage is not suitable for long-term holding of mum flowers.

5.4.6 *Subatmospheric Pressure (Hypobaric) Storage*

Cut chrysanthemums can be stored for up to 21–28 days without any loss in the postharvest qualities [4]. However, the economic aspects should be taken into consideration before advocating this method for commercial application.

5.4.7 *Floral Preservative and Bud-Opening Solutions*

The use of floral preservative is beneficial at all stages in the marketing channel. The type of preservative solution is very critical in cut mums. Several compounds, either alone or in combination, have been tried for their effectiveness in extending the vase life of cut flowers. Among these compounds, only few are found to be useful for chrysanthemum. The life of cut mums was extended by 8 to 10 days by dipping the flower stems in 1200 to 4800 ppm of silver nitrate for a very short time (5 s). A 10-min dip of flower stems in 1000 ppm of silver thiosulfate was found to be equally effective [81]. Since the silver ion from silver nitrate is relatively immobile in plant tissues, silver thiosulfate has been found to be very effective [85]. STS treatment cannot replace the preservatives, but enhances their effect.

Most of the commercially available preservative solutions contain sugar and 8-HQC [3]. Sodium dichloro-s-thiazine-trione (SDT) plus sugar was as effective as silver ions and sucrose or other preservative solutions for extending the vase life in chrysanthemums [82, 83]. However, SDT without sugar was not effective. These solutions maintain high rates of water absorption in cut flowers by preventing stem plugging. The effectiveness of any floral preservative solution depends upon the time lapse between harvesting and placing the stems in the solution. Flowers that are allowed to wilt in the field after harvesting may not regain turgour after they are returned to water or preservative solution. Recutting of the stem just before placing in water or cutting in water is beneficial for faster uptake of water and recovery from wilting [4]. Since salt levels of about 100 ppm or more in the water can decrease the lasting qualities of cut mums, the water used in holding solution should be low in soluble salts [4].

Preconditioning of cut mums by pulsing for 16 to 20 h in a solution of 5% sucrose, plus 200 ppm Physan-20, improves the storageability and ability of the flowers to withstand transportation [86]. These treatments also extend the post-storage or post-shipping vase life of cut flowers.

Mums cut in bud stages can be opened to their normal size using bud opening solution. The stage of development of the bud is critical in bud opening [84]. Chrysanthemums harvested in the tight bud stages can be made to open by using chemical preservative solution. Buds of standard mums open well in a solution of 8-HQC (200 ppm) plus 2% sucrose [62]. This solution can also be used as a holding solution. The vase life of bud-opened flowers after dry storage of buds was equal to that of freshly harvested flowers. The quality of water used in the bud-opening solution is critical. Kofranek et al. [85], recommended the use of distilled or deionized water to prepare the bud-opening solution. They used 25 ppm silver nitrate, 75 ppm citric acid and 2% sucrose. The cultivars vary in their response to sucrose concentration. The bud-opening solutions should be used continuously throughout the opening process to obtain high quality flowers. Staby et al. [4] reported that use of opening solution only for 3 to 7 days after harvest resulted in "flat" blooms. Foliar chlorosis during storage can be prevented by providing continuous light or by preharvest spray of antisenescence agents like N^{6-}-benzyl adenine, which also delay leaf abscision in the storage environment [86].

5.4.8 Control of Pests and Diseases

Maintaining proper hygienic conditions in the growing as well as postharvest environment is essential for reducing damage due to insect pests and diseases. Diseases like *Botrytis* are very difficult to control in storage environment. Integrated efforts to control pests and diseases and the use of other improved loss reduction biotechnologies will aid in delivering flowers that have the best qualities and last longer in the home.

References

1. Mathis R (1981) L'évolution de la culture du chrysanthème en Europe. Acta Hortic 125:13
2. Machin B, Schopes N (1978) Chrysanthemums Year round Growing. Blandford Poole Dorest UK, p 233
3. Laurie A, Kiplinger DC, Nelson KS (1968) Commercial flower forcing. McGraw-Hill, New York USA
4. Staby GL, Robertson JL, Kiplinger DC, Conover CA (1978) Chain of life. Ohio Flor Assoc Bull p 71
5. Krause J, Lisieck A (1981) Chrysanthemum cultivars grown in the plastic house. Acta Hortic 125:149
6. Mastalerz JW (1977) Greenhouse Environment. Wiley S, New York USA
7. Carpenter WJ (1964) Response of snapdragons and chrysanthemums to supplemental reflective sunlight. Proc Am Soc Hortic Sci 84:626
8. Doorenbos J, Kofranek AM (1953) Inflorescence initiation and development in an early and late chrysanthemum varieties. Proc Am Soc Hortic Sci 61:555
9. Cathey HM (1954a) Chrysanthemum temperature study. A. Thermal induction of stock plants of *Chrysanthemum morifolium*. Proc Am Soc Hortic Sci 64:483
10. Cathey HM (1954b) Chrysanthemum temperature study. B. Thermal modifications of photoperiod previous to and after flower bud initiation. Proc Am Soc Hortic Sci 611:492

11. Cathey HM (1954c) Chrysanthemum temperature study. C. The effect of night, day and mean temperatures upon the flowering of *Chrysanthemum morifolium*. Proc Am Soc Hortic Sci 64:499
12. Cathey HM (1957) Chrysanthemum temperature study. F. The effects of temperature on critical photoperiod necessary for initiation and development of flowers of *Chrysanthemum morifolium*. Proc Am Soc Hortic Sci 69:485
13. Ben-Jsacov B, Langhans RW (1969) After-lighting of chrysanthemums. NY State Flower Grow Assoc Bull 285:1–3
14. Post K (1950) Controlled photoperiod and spray formation of chrysanthemums. Proc Am Soc Hortic Sci 55:467
15. Post K, Lacy DB (1951) Interrupted short day improves standard chrysanthemums. NY State Flower Grow Bull 70:2
16. Kawata J, Tayoda T (1981) The response to photoperiod and temperature in Japanese July to September flowering chrysanthemums. Acta Hortic 125:93
17. Miller RO, Kiplinger DC (1962) The effect of day length and temperature on the production of tubular florets (quills) by Indianapolis chrysanthemum. Ohio Agric Exp Sta Res Cir 109
18. Koths JS, Auzina R (1965) Summary of 1964/65 CO_2 studies with chrysanthemums. Conn Flor Newsletter (Exten Serv) July:3–6
19. Mastalerz JW (1977) Response of chrysanthemum cultivars to CO_2 enrichment at 1000 ppm. In: Mastalerz JW, Greenhouse Environment. Wiley New York USA 296
20. Barrett RE, Ormood DP, Jung R (1978) Soil heating effects on bench grown chrysanthemum. HortScience 13:591
21. Brown WW, Ormond DP (1979) Response of Chrysanthemum to soil heating. Sci Hortic (Amst) 13:67
22. Mostafa AT, Morgan JV (1981a) Root zone warming of spray chrysanthemum in hydroponics. Acta Hortic 115:217
23. Mostafa AT, Morgan JV (1981b) The effects of root zone warming and low night temperature on the growth and quality of spray chrysanthemum in nutrient solution culture. Acta Hortic 125:133
24. Morgan JV, Haire RO (1978) Heated hydroponic solutions as an energy saving technique. Acta Hortic 76:173
25. Cockshull KE (1975) Premature budding in year round chrysanthemum. Rep Glasshouse Crops Res Inst 128
26. Cockshull KE (1976) Flower and leaf initiation by *Chrysanthemum morifolium* Ramat in long days. J Hort Sci 51:441
27. Cockshull KE (1979) Effects of irradiance and temperature on flowering of *Chrysanthemum morifolium* Ramat in continuous light. Ann Bot 44:451
28. Hughes AP, Cockshull KF (1971) The effects of light intensity and carbon dioxide concentration on the growth of *Chrysanthemum morifolium* cv. Bright Golden Anne. Ann Bot 35:899
29. Hughes AP, Cockshull KE (1972) Further effects of light intensity, carbon dioxide concentration and day temperature on growth of *Chrysanthemum morifolium* cv. Bright Golden Anne in controlled environments. Ann Bot 36:533
30. Krahn JN (1968) Cut flower transportation and handling. Agriculture and Forestry Bull 1(4):10
31. Blandon F (1981) La relation des exigences in facteurs externes pour la floraison des chrysanthemums. Un example. *Chrysanthemum morifolium* 'Shuokan'. Acta Hortic 125:119
32. Votruba R (1981) Premature budding in *Chrysanthemum morifolium* in relation to the age of the stock plants. Acta Hortic 125:111
33. Furuta T, Kiplinger DC (1955) Chronological age of cuttings, a factor influencing the spray formation of pompom chrysanthemum. Proc Am Soc Hortic Sci 3:83
34. Kofranek AM (1981) Affecting spray of pompom chrysanthemums during natural season flowering. Acta Hortic 125:167
35. Runger W (1976) Licht und Temperatur im Zierpflanzenbau. Paul Parey, Berlin und Hamburg 191
36. Cox R (1969) Quailling in chrysanthemums. Plant Disease Reporter 53:518

37. Chan AP (1950) The development of crown and terminal flower buds of *Chrysanthemum morifolium*. Proc Am Soc Hortic Sci 55:461
38. Dirkse FB (1981) Preharvest treatment of chrysanthemum against *Botrytis cineraria*. Acta Hortic 221
39. Ovlikowski LB, Wojdyla A (1981) Chemical control of chrysanthemum white rust. Acta Hortic 125:201
40. Grouet D, Montfort F, Milley D (1981) Traitment de la rouille blanche du chrysantheme pur thermotherapie. Acta Hortic 125:193
41. Monsion M, Macquire G, Bachelier JC, Danez J (1981) Recent progress in chrysanthemum viroid detection. Acta Hortic 125:227
42. Favier AJ, Roux J, Faivre ML (1981) Biological control of *Agrobacterium tumefaciens* (Schmit et Townsend) Conn. on Chrysanthemum with K84, *Agrobacterium radiobactor* (Beijerinck et van Delder) Conn. strain. Acta Hortic 125:233
43. Vrie M, Van de Dilkse F (1981) Biology and control of the leafminer (*Liriomyza trifoli*) (Burgess) on glasshouse chrysanthemum. Acta Hortic 125:257
44. Friman ID, Marlin PH (1968) White rust of chrysanthemum. Ann Appl Biol 62:429
45. Dickens SW (1970) Infection of chrysanthemum flowers by white rust *Puccinia horiana*. Pl Pathology 19:122
46. Post K (1947a) Year-round chrysanthemum production. Proc Am Soc Hortic Sci 49:417
47. Post K (1947b) Chrysanthemum troubles. NY State Flower Grow Bull 27:44
48. Post K (1942) Effects of day length and temperature on growth and flowering of some florist crops. Cornell Agric Expt Sta Bull 787
49. Post K (1948b) Day length and flower bud development in *Chrysanthemums*. Proc Am Soc Hortic Sci 51:590
50. Post K, Kamemoto H (1950) Photoperiod and bud formation on chrysanthemum cuttings. Proc Am Soc Hortic Sci 55:483
51. Post K (1949) Florist crop production and marketing. Orange Pub Co, New York
52. Kiplinger DC, Alger J (1948) Interrupted shading of chrysanthemums. Proc Am Soc Hortic Sci 52:478
53. Cathey HM (1947) Participation of phytochrome in regulating internode elongation of *Chrysanthemum morifolium* (Ramat.) Hemsl. J Am Soc Hortic Sci 99:17
54. Canham AE (1975) Supplementary lighting for spray chrysanthemum. Acta Hortic 51:253
55. Miske TH (1981) Chrysanthemum. Kuv Z Kultar Gb–Gw, 4:74
56. Jerzy M (1981) Long day glasshouse – short day glasshouse system in year-round chrysanthemum production. Acta Hortic 125:143
57. Jergy M (1979) Light factor control in the year round programme of the cultivation of chrysanthemum. Zesz Nauk ATr W Bydgoszczy Rolnictwo 7:1
58. Challet P (1981) Contribution to the technological improvement of chrysanthemum crops for All Saints Day in France. Creation and utilization of anticipated photoperiodical techniques. Orchestra method. Acta Hortic 125:127
59. Marousky FJ (1973) Recent advances in opening bud cut chrysanthemum flowers. HortScience 8:199
60. Kofranek AM (1972) Decreasing water loss of cut chrysanthemum buds. J Am Soc Hortic Sci 97:578
61. Byrne TG, Lert PJ (1972) Single cut harvesting of chrysanthemums for commercial marketing. Ca Agriculture 26(1):8
62. Marousky FJ (1971) Handling and opening bud cut chrysanthemum flowers with 8-hydroxy quinoline citrate and sucrose. Marketing Res Rep USDA 905
63. Marousky FJ (1972) Water relations effect of floral preservatives on bud opening and keeping quality of cut flowers. HortScience 7:114
64. Marousky FJ, Nanney J (1971) Bud cut flowers concept for once-over harvest of chrysanthemum. Proc Fla Sta Hortic Soc 84:399
65. Waters WE, Conover CA (1969) Chrysanthemum production in Florida. Fla Agric Expt Stn Bull 730
66. Marousky FJ (1971b) Incidence and control of foliar chlorosis on bud cut chrysanthemum. HortScience 6:298

67. Boodley JW, Sheldrake R Jr (1973) Boron deficiency and petal necrosis of Indianapolis white chrysanthemum. HortScience 8(1):24
68. Criley RA, Carlson WI (1970) Tissue analysis standards for various floricultural crops. Flor Rev 145(3771):19
69. Gogue GJ, Sanderson KC (1973) Boron toxicity in chrysanthemum. HortScience 8(6):473
70. Lunt OR, Kofranek AM, Oertti JS (1963) Deficiency and symptoms and mineral nutrient levels in Good News chrysanthemum. Flor Rev 132(3430):13
71. Woltz SS (1956) Studies on the nutritional requirements of chrysanthemums. Proc Fla Sta Hortic Soc 69:352
72. Wiersama O, Boyer WC (1971) Vacuumkoloelen EN transport van snijbloemen. Sprenger Institut Rapport 1808:8
73. Reist A (1978) A comparison of wrapping material for cut flowers. Rev suisse vitic Arboric Hortic 10(5):245
74. Jong J de (1978) Color formation in chrysanthemum. Vakbl Bloemisterij 33(31):26
75. Wiersma O, Boer WC (1972) Transport temperature en vacuumkoelen van snijbloemen (Westland 1972). Springer Inst Rep 1847:17
76. Thornton NC (1930) The use of carbon dioxide for prolonging the life of cut flowers. Amer J Bot 17:614
77. Longley LE (1934) Some effects of storage of flowers in various gases at low temperatures on their keeping quality. Proc Am Soc Hortic Sci 87:521
78. Neff MS (1942) Effects of storage conditions on cut roses. Bot Gaz 103:794
79. Tingo JA (1956) The effect of modified storage at low temperature and treatments after low temperature storage which affect keeping quality of cut flowers. PhD Diss, Cornell Univ USA
80. Haugue A, Bryant W, Laurie A (1947) Prepackaging of cut flowers. Proc Am Soc Hortic Sci 49:427
81. Farnham DS (1947) Silver impregnated stems aid chrysanthemum flower longevity. Calif Agric Ext Flower and Nurs Rep July, 1–5
82. Kofranek AM, Kohl HC, Kabota J (1974) A slow release chlorine compound as a vase water additive. Flori Rev 154(4000):21, 63
83. Kofranek AM, Kohl HC, Kabota J (1975) Stabilized chlorine compound as vase water additive. Calif Flower and Nurs Reb Feb, 5–7
84. Kofranek AM, Halevy AH (1972) Conditions for opening of cut chrysanthemum flower buds. J Am Soc Hortic Sci 97:578
85. Kofranek AM, Halevy AH, Kabota J (1972) Bud opening of Albatross and Shoesmith mums. Flor Rev 149(386):27
86. Kofranek AM, Halevy AH (1981) Chemical pretreatment of chrysanthemum before shipment. Acta Hortic 113:89

Chapter 6 Tulip

6.1 Introduction

Tulip is regarded as the most attractive spring bulb flower. In recent years the popularity of tulips has suffered to a considerable extent because of the short keeping quality of the flower. Tulips are grown for cut flowers or as potted plants for Valentine's Day, Easter and Mother's Day [1].

The quality attributes in tulip are either quantitative (flower size, stem length, vase life and length of flower) or qualitative in nature (shape of flowers, freedom from white-tip perianth, green perianth, abnormal anthers, flower blasting and injuries due to pests and diseases) [2]. A minimum room temperature vase life of 5 to 6 days is required for commercial cultivars used for cut flower production [2]. A top grade cut tulip flower must have a stem strong enough to hold the weight of flowers without bending or drooping, colour and size representative of the cultivar, and have minimum postharvest vase life of 5 to 6 days at room temperature. Senescence of the perianth is considered to be a good criterion for monitoring the useful life of cut tulips.

Flower bulbs such as tulips, hyacinth and other hardy bulbs were, in the past, restricted to Holland, with very little production taking place in other parts of the world. However, in the recent past several countries such as the USA, Denmark, the UK etc. have sizeable tulip production [1].

6.2 Physiology and Biochemistry

6.2.1 Growth and Development

Since the 1920's, there have been many investigations on the physiology of floral development of bulbous and tuberous crops [3, 6]. However, detailed description of the growth and development of tulip plant for the whole life cycle is rare.

The relative growth rate (RGR) and the quality of flowering shoot is influenced by photoperiod, temperature and nutritional status during the storage of bulbs, as well as during the greenhouse forcing period [4, 5, 7, 21]. While studying the effects of photoperiod on the growth of daughter bulbs of Rose Copland tulips, Rees [6] demonstrated that long days promote faster shoot extention. Hanks and Rees [11, 12] observed the effects of long days on shoot extension within 1 – 2 weeks of the start of day length treatments. These effects were found to be due to an increased elongation of all internodes and to persist through the flowering period to the cessation of growth.

The response of plants to photoperiod to a large extent depends upon the growing temperature. In the case of tulip, Hanks and Rees [11] demonstrated that plants grown at 9 °C responded better to long day treatment than those grown at 18 °C. In all the five tulip cultivars (Apeldoorn, Golden Apeldoorn, Paul Richter, Rose Copland, and Trance) studied by Hanks and Rees [11], the stem length was shortest under SD (8 h day) treatment and increased progressively when the day lengths were increased to 12 and 16 h. Other flower attributes such as flowering date, flower size and keeping quality were not influenced by long day treatments.

Flowering in tulip can be greatly advanced by using prepared or precooled bulbs [13–16]. The flower size was found directly correlated with the duration of precooling of bulbs (Fig. 6.1). Incomplete precooling of bulbs causes production of flowers before the complete opening of leaves, short flower stems, and unmarketable flowers [13–15]. Low temperature was found to induce both qualitative and quantitative changes in endogenous hormones, particularly gibberellins [17–19]. Exogenous application of gibberellins was found to have synergetic effects to precooling and can partially substitute for precooling [8, 19, 20]. Although all the treated plants flowered earlier, the order of effectiveness of various gibberellic acids was: $GA_7 > GA_4 > GA_1 > GA_3 > GA_{13} > GA_5 > GA_3$ [20]. The GA_7-treated bulbs flowered 24 days earlier than fully cooled bulbs (not treated). Similarly, GA_4 also hastened the flowering by 33 days, increased the first internodal length and reduced the length of the last internode. The occurrence of white tipping in flowers was also reduced with GA treatment.

Normal growth of flower bud occurs only when the bulbs are cooled at the right stage and time [22–24, 30]. For early forcing, the bulbs should be removed to low temperatures only when the flower bud differentiation (G stage) has been completed. High temperature (30 °C) treatment applied to the bulbs immediately after lifting induces rapid flower bud differentiation and subsequent flower growth. In the case of Apeldoorn tulips, maintaining the bulbs at 30 °C immediately after lifting for 5 weeks and then at 20 °C until the completion of the G stage, followed by regular 12-week cold treatment at 5 °C,

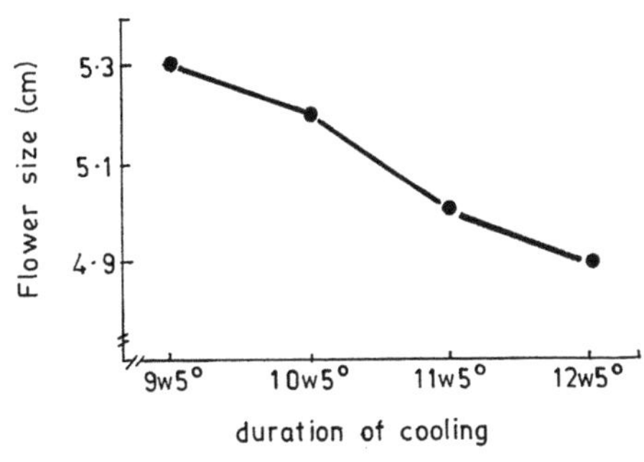

Fig. 6.1. Flower size as influenced by the duration of low temperature storage of bulbs [32]

Table 6.1. Effects of root temperature during forcing phase on quality of Apeldoorn tulips [36]

Root temperature (°C)	Root length after 22 days (mm)	Days to anthesis	Keeping quality of flower in	
			Water	Tulip Krislite[a]
12	15.8	25	9	10
15	20.0	24	9	10
18	17.1	24	9	7
21	12.8	21	9	7
24	17.0	19	9	7

[a] A commercial floral preservative

caused faster rooting, better development of roots, early flowering and heavier aerial parts [24].

Root temperature during forcing influences the root length, time taken to flower anthesis, and keeping quality of cut tulips. Le Nard [24] suggested temperatures less than 18 °C for producing flowers with best lasting qualities (Table 6.1).

Irrigation water quality and calcium levels in the growing medium have great influence on flower quality and production of blind shoots (or plants without flowers). Calcium requirement of flower primordia during the development has been found to be very high [36] (11 mg/plant). Since calcium is relatively immobile in the plant, its continuous supply through soil is essential in order to prevent blindness in tulips [25]. The nitrate form was found to be the best source of calcium, while the sulfate form produced very poor response [26]. Low nitrate levels and higher root temperatures were also found to increase topple disease [26].

6.2.2 Senescence of Cut Tulip

The attractiveness of cut tulips can be drastically affected by floral stem elongation and flower senescence [27, 28]. Senescence of the perianth is regarded as the best criterion in measuring the useful life of cut tulips [28]. An ageing perianth passes through four distinct developmental processes, viz., stem elongation, thermostatic movement and elongation of perianth, and colour fading [29]. These processes, when accelerated, cause drastic reduction in the useful life and quality of flowers.

Excessive postharvest stem elongation results in stem bending, thereby detracting from the appearance of cut flowers well in advance of senescence of the perianth. The internode just below the flower has been shown to elongate faster than the rest, accounting for nearly 43% to 77% of the total postharvest stem elongation [29]. This elongation is shown to be regulated by hormonal balance in the flower stalk. Exogenous application of indole acetic acid or naphthalene acetic acid to deblossomed shoots causes similar stem elongation. Since the removal of floral parts inhibited stem elongation, it is commonly believed that the primary control site for this response exists in the flower bud. Among different floral structures, the perianth has been shown to be the primary site, with gynoeceum and androceum as secondary centres of control. In contrast, removal of leaves promoted stem elongation by making available higher amounts of native auxins in flower stalks [29].

Exposure of flowers to ethylene can cause drastic reduction in postharvest lasting qualities of cut tulips by stimulating senescence of the perianth [28]. Prevailing temperature conditions also influence different developmental processes in ageing flowers. In addition to temperature conditions in the cut flower holding room, greenhouse forcing temperatures also affect postharvest longevity of cut tulips. Yokoi et al. [30] found that vase life of cut William Pitt tulips was increased by 2 to 5 days when the plants were forced at 10 °C, as compared to those from plants grown at 17.5 and 25 °C.

6.3 Postharvest Losses in Cut Tulips

The reduction in marketable cut tulips is associated mainly with bud blasting, "toppling" or with mechanical damage resulting from faulty handling practices in the postharvest phase. Furthermore, excessive elongation of the topmost internode (immediately below the flower) causes bending of the stalk due to weight of the flower, thereby reducing its aesthetic value. Physiological wilting resulting from the inability of cut flowers to rehydrate on arrival in the retail florists' shop add greatly to postharvest loss.

Cultivar differences in respect to flower longevity are well recognized. Most of the Darwin hybrids like Apeldoorn, General Eisenhower, Gudoshnik, London, Oxford and President Kennedy are generally regarded as poor-lasting cut tulips. These last on an average 3 to 4 days when maintained at 21.5 °C, as against the minimum acceptable standard for postharvest longevity of 5 to 6 days for cut tulips.

By proper cultivar selection, the postharvest longevity of tulip flowers can be increased by almost 100% or more. Since shelf life of 5 to 6 days is considered a minimum standard for commercial cut tulips, any cultivars having shorter shelf life than this are unsuitable for forcing for cut flowers. In this regard, most of the Darwin hybrids have been shown to have very short vase life. This is partly due to the fact that they differ in their sensitivity to ethylene in the storage and forcing environments [31].

By manipulating pre-harvest growing conditions, it is possible to improve the quality of cut tulips [32]. Several scientists have suggested the possibility of regulating the flowering and quality of flowers by modifying the temperature and photoperiod in the greenhouse.

Stage of harvesting is another factor that influences the postharvest lasting qualities in cut tulips. The best stage for harvest is when the flowers are in green bud stage or when the upper of half portion of the flower is showing colour. Flowers harvested at this stage have been shown to be more tolerant to mechanical injuries in transit, to last longer and also to store better. Flowers harvested with the bulb attached can be stored for a longer period of time than those without bulbs [28].

The benefits of the use of floral preservatives in tulip are variable. In general, cut tulips have been shown to respond poorly to floral preservative solutions or chemical treatments. The preservative solutions containing sucrose cause excessive stem elongation and lead to a drastic reduction in the ability of stems to hold the flower in upright position. It is important to realize that most rapid water or solution uptake takes place in the first 24 h of placing in holding solution. Recutting of stems before placing them in water can greatly improve uptake of water by flower stems.

Excessive postharvest elongation of flower stalks in the postharvest stage results in bending of stems well before senescence of the perianth [33]. Since this elongation is regulated by the hormonal balance in the cut flower, several research workers have demonstrated the use of growth retardants like Ancymidol [34] and Ethephon [35] to control postharvest elongation of the uppermost internode and flower stalk as a whole. Einert [34] demonstrated that vase solutions containing 25 ppm Ancymidol significantly reduced the elongation of the last internode in cut tulips without adversely affecting flower quality or vase life. Pre-harvest spray of 50 ppm of Ancymidol 24 to 72 h before harvest was also equally effective in reducing postharvest stem elongation in Hibernia and Paul Richter tulips. However, water uptake by cut stems was slightly reduced in the case of cut Hibernia flowers that were treated with 50 ppm Ancymidol before harvest.

References

1. De Hertough AA (1980) Bulbous plants. In: Larson RA (ed) Introduction to Floriculture. Academic Press, London, 215
2. De Hertough AA (1977) Holland Bulb Forcers' Guide. Netherland Flower Bulb Institute, New York
3. Hartsema AM (1961) Influence of temperatures on flower formation and flowering of bulbous and tuberous plants. In: Rhuland W (ed) Handbuch der Pflanzenphysiologie. Springer-Verlag, Berlin 16:123
4. Gilford JMD, Rees AR (1973) Growth of tulip shoots. Hortic (Amst) 1:143
5. Open den Kelder P, Benschop M, De Hertough AA (1971) Factors affecting flower stalk elongation of flowering tulip. J Am Soc Hortic Sci 96:603
6. Rees AR (1972) Growth of Bulbs. Academic Press, London
7. Rees AR, Turquand ED, Briggs JB (1972) Interrelations of bulb storage treatment and housing date on flowering date, stem length and flower differentiation of tulip. Expt Hort 23:52
8. Shoub J, De Hertough AA (1974) Effects of ancymidol and gibberellin A_3 and A_4+7 on *Tulipa gesneriana* L. cv. Paul Richter during greenhouse development. Sci Hortic (Amst) 2:55
9. Hansen H (1976) Studies on the growth pattern *Tulip* Tidsskrift for Planteavl 80:869
10. Rees AR (1977) The cold treatment of tulip cultivars. Sci Hortic (Amst) 7:387
11. Hanks GR, Rees AR (1979) Photoperiod and tulip growth. J Hortic Sci 54:39
12. Hanks GR, Rees AR (1980) Day length and flowering performance of tulips. Acta Hortic 109:177
13. Cocozza TM (1972a) Richerchhe sugli effetti della vernalizzazione dei bulbi sullanticipo di fioritura e sulle caratteristiche qualitative del tulipano coltivato in pien aria e sotto tunnel. Inf Bot It 3:213

14. Cocozza TM (1972b) Effect: della vernalizzazione dei bulbi sulla fioritura e sulle caratteristiche qualitative del tulipano, narciso, e giacinto. Culture protette 1:31
15. Cocozza TM (1974) Recherche sur les effets de la refrigeration des bulbes et de la couverture par tunnel sur la floraison et les caracteristiques qualitative de la tulipe. Acta Hortic 43:53
16. Cocozza TM (1976) Utilizzazione delle basse temperature in floricultura. Il freddo 4:195
17. Cocozza TM, Caputo V (1980) A research on the possibility of using gibberellin for tulip forcing. Acta Hortic 109:163
18. De Hertough AA (1974) Principles of bulb for hyacinth, daffodils, easter lilies and dutch irises. Sci Hortic (Amst) 2:313
19. Rees AR, Hanks GR (1979) Potential uses of plant growth regulators in bulb growing out forcing. Acta Hortic 91
20. Hanks GR, Rees AR (1980) Partially cooled tulips. Response to GA_3 and other gibberellins. Acta Hortic 109:169
21. Moe R, Wickstorm A (1973) The effect of storage temperature on shoot growth, flowering and carbohydrate metabolism in tulip bulbs. Physiol Plant 28:81
22. Kawata J (1975) Optimum temperature and duration of low temperature treatment for forcing tulips. Acta Hortic 47:201
23. Le Nard M, Canat J (1968) Influence des temperatures de conservation des bulbes sur l'elongation la floraison et ala bulbification de la tulipe (*Tulipa gesneeriana* L.). Ann Amelior Plantes 18(2):181
24. Le Nard M (1980) Effects of duration of high temperature treatment on subsequent flower differentiation, rooting and flowering of tulip bulbs. Acta Hortic 109:65
25. Bjirggaard A (1975) Drivning of blomsterlog i vandbede. Gartner Tidende, 11
26. Cheal WF, Hewitt EJ (1964) The effects of rates of supply of nitrogen, phosphorus, potassium, and magnesium on leaf and stem growth, flowering and topple of Golden Harvest and Elmus tulips. Ann Appl Biol 53:477
27. Hanks GR, Rees AR (1977) Stem elongation in tulip and narcissus: the influence of floral organs and growth regulators. New Phytiol 78:579
28. Staby GL, Robertson JL, Kiplinger DC, Conover CA (1978) Chain of Life. Ohio Flor Assoc Bull pp 71
29. Benschop M, De Hertough AA (1971) Postharvest development of cut tulip flowers. In: First Int Symp Flower bulbs, Noordiwijk Lisse. Acta Hortic 23:121
30. Yokoi M, Kosugi K, Shinoda K (1977) Cut flower keeping quality as influenced by growing and storage temperatures. Tech Bul Fac Hort Chiba Univ 25
31. De Hertough AA, Dilley DR, Blakely B (1980) Response variation of tulip cultivars to exogenous ethylene. Acta Hortic 109:205
32. Munk WJ de, Kamerbeek GA (1975) The effects of preharvest conditions on the quality of the cut tulips. Acta Hortic 41:69
33. Einert AE (1971) Reduction of last internode elongation of cut tulips by growth retardants. HortScience 6(1):459
34. Einert AE (1975) Effects of ancymidol on vase behaviour of cut tulips. Acta Hortic 42:97
35. Moe R (1980) The use of ethephon for control of plant height in daffodils and tulips. Acta Hortic 109:197
36. Klougurt A (1980) Calcium uptake of tulips during forcing. Acta Hortic 109:89
37. Benschop M, De Hertough AA (1971) Postharvest development of cut tulips flowers. Tech Communications of Int Soc Hortic Sci 23:121
38. Munk WJ de, Gizjenberg J (1977) Flower-bud blasting in tulip plants mediated by the hormonal status of the plant. Sci Hortic (Amst)7:255
39. Moe R (1979) Hormonal control of flower blasting in tulip. Acta Hortic 91:221
40. Ruduicki RM, Nowak J, Saniewski M (1976) Effect of gibberellic acid on sprouting and flowering of some tulip cultivars. Sci Hortic (Amst) 4:387

Chapter 7 Orchid

7.1 Introduction

Orchids are considered to be the very special cut flower in the world. They belong to a largest family in the plant kingdom, with over 20000 known species [1]. The terrestial forms are able to produce the photosynthates needed for their growth and flowering. Out of all the existing orchid types, only a few genera, such as *Cattleya*, *Phalaenopsis*, *Dendrobium*, *Cymbidium* and *Vanda*, are commercially grown for cut flowers [3]. *Cymbidium* and *Vanda* are also produced in the open in subtropical areas. These are regarded as a special occasion flowers, replacing roses in weddings and other formal social functions, Easter and Mother's Day celebrations. Orchids are in demand throughout the year and make very good specimen flowers for transworld delivery.

Orchids are classified into monopodial and sympodial types. In monopodial types (e.g. *Phalaenopsis polyrrhiza*), the main stem continues to grow indefinitely and lateral branches do not become the part of the main axis. The flowers are produced in the leaf axils.

The sympodial orchids (e.g. *Cattleya, Vanilla, Dendrobium* and *Cymbidium*) have a prostate rhizome whose growth terminates with an upright pseudobulb, leaf and flowers. The growth of each shoot stops usually after one or rarely two or more seasons. The development of the plant is continued by the growth of buds in the axils of scaly leaves at the base of the parent shoot. In sympodial orchids with perennial stems or climbing branches, the branches arise from the axils of stem leaves (*Vanilla*), which are comparable to the monopodial orchid, but all the branches, including the main stem, are terminated by inflorescence [2].

7.1.1 Postharvest Quality Requirements

Since orchid flowers are the "special occasion" flower, the condition of the flower is very important. Furthermore, if a single flower in a consignment is beginning to fade or destined to fade soon, and if the ethylene liberated by it is not eliminated or inactivated, the whole consignment will fade in transit. Therefore, it is imperative that only sound flowers (unfaded, unpollinated, with intact pollinia) packaged in a clean atmosphere devoid of ethylene [3].

The turgidity of the flower, the colour, the shape or the form of the labellum and other structures in the flower are more important in marketability of orchid flowers. In general, the flowers with a very special type of pigmentation or unusual form are in more demand on the international market. The longevity of the flower is another factor which makes orchids very popular. Undisturbed and unpollinated flowers last upt to 3 weeks or more. However, the postharvest-lasting quality is secondary to other factors, like condition or freshness, shape (or form), and quality of flowers [3].

7.1.2 Species and Commercial Cultivars

Orchid culture has become the most important hobby for countless thousands and a "big business" for a number of people in the world. As a result of consistent efforts by thousands of orchidists, today we have over 20000 species in 750 genera known throughout the world. In addition to these natural species, there are nearly 43000 hybrids which are listed as existing in the world [1, 2].

Out of the huge list of known genera of orchids, only a few are commercially grown for cut flowers. The present-day species under cultivation are mainly hybrids belonging to different genera. The *Cymbidium* are very popular in Europe, while others, like *Paphiopedilum*, *Odontoglossom*, *Cattleya*, are also in particular demand. In the USA the *Cattleya*, *Phalaenopsis*, *Cymbidium*, and *Vanda* are grown in the greenhouses, while *Cymbidium* and *Vanda* are also produced outdoors [2]. In recent years, the demand for *Dendrobium, Oncidium, Laelia* and *Epidendrum* has increased considerably.

7.1.3 Production

Orchids grow all over the world except for polar regions and deserts. Most of the known species prefer tropical and sub-tropical environment. From the florist's point of view, the genera like *Cattleya, Cymbidium, Vanda,* and *Phalaenopsis* are of great value.

The major orchid-producing centres of the world are the USA (Hawaii, California and Florida), the UK, Holland and some of the African and Asiatic countries. In Hawaii, orchids, mainly *Vanda*, rank second only to Anthuriums [4, 10]. Although the out-of-state sale of orchids varies greatly from year to year, the production and sale of orchid flowers has increased steadily.

7.2 Physiology and Biochemistry

7.2.1 Growth and Development

Flowering in orchids varies both in time and number of flushes within a year from one genus to the other. Flowering in *Cattleya* is controlled by day length and/or temperature, whereas the flower bud initiation in *Cymbidium* is regulated by temperature only [2, 6]. Apart from the photoperiod, the intensity of light during the growth of plants is also important. Orchids grow best in light intensities from 1000 to 3000 ft candles. At higher light intensities, the leaves turn yellow and develop necrosis, while at lower light intensities the development of the plant is delayed.

The length of time required for flower bud initiation and flower development varies with species and cultivar. The flowering in cultivars belonging to the genus *Cattleya* is promoted by short days [5, 6]. In general, the flowering in these varieties occurs 90 to 120 days after the start of short days. Therefore, lighting should be discontinued approximately 17 weeks before the time of desired flowering. In contrast, flowering in *Cymbidium* is unaffected by photoperiod [5].

Although the flowering in several *Cattleya* is governed by light; temperature also plays a major role in flower bud development, as well as initiation, in some cultivars. The optimum temperature for growing *Cattleya* is 15.6 °C (2.8 °C lower during flower formation in thermosensitive varieties). Low temperatures (below 12.7 °C) during flower bud development result in deformed flowers. *Phalaenopsis* are best grown at 18.4 °C, while *Cymbidium* prefers cool weather (10.6 to 12.7 °C). Flowering in *Cymbidium* is promoted by cooler temperature (6.8 to 12.7 °C) [2].

7.2.2 Senescence of Cut Flowers of Orchids

The blossoms of orchid fade normally in senescence, but premature fading may be induced by subjecting the flowers to certain gases, by pollination, and disturbance or removal of pollinia [7, 8]. Premature fading also occurs when flowers are exposed to ethylene gas or if the blossoms are stored with plant materials (ripening fruits) that produce ethylene [10]. Lindner [11] demonstrated that ethylene levels of 1 ppm caused the normal blooms to fade prematurely, while Davidson [12] was able to produce the same effects by exposing *Cymbidium* flowers to 0.0002 ppm ethylene for 24 h or 0.1 ppm for 8 h.

From the results of biological assays, Fischer [13] demonstrated the production of ethylene by blossoms of several species of orchids (*Dendrobium* and *Phalaenopsis* spp.). Young tomato and marigold plants sealed in a container with fading *Vanda* orchids displayed epinasty, a typical ethylene response. This also confirms the findings of Fischer. Lindner [11] suggested that ethylene produced by the blossoms of *Vanda* causes their own fading. The fading *Vanda* flowers were shown to produce 3400 μl of ethylene per kg per hour, which is many hundred times more than the level needed for causing premature fading of blossoms.

Ethylene injury in orchids is characterized by a progressive drying and bleaching of the sepals, beginning at the tip and extending towards the base [12]. The rate of respiration increases as the flowers develop from the bud stage to anthesis. Orchid flowers harvested in the tight bud stage had lower rates of respiration than open flowers [14]. Once the flowers were fully open, the rate of respiration was lower than it was for flowers harvested in the tight bud or at anthesis stage. Respiration rates declined continuously during the postharvest period until the flower faded [14].

Arditti and his coworkers [15–22] studied the post-pollination phenomenon in *Cymbidium* orchids. Premature fading of flowers due to pollination is characterized by the formation of anthocyanin in both gynostemia (column) and labellum (lips), swelling of column, stigmatic closure, and wilting of sepals and petals. Anthocyanin levels begin to decrease with age [15, 16]. Application of NAA to *Cymbidium* stigmas also induces anthocyanin formation in column and lips and all the other symptoms of post-pollination behaviour of orchid flowers [20]. Anthocyanin formation and wilting, but not swelling of column and stigmatic closure, can be inhibited by actinomycin D, ethionine and puromycin, suggesting the need for the *de novo* synthesis of DNA-dependent RNA, and the production of new proteins [20]. Swelling of column and stigmatic closure depends upon new synthesis and is brought about by the activation of pre-existing compounds. Removal of the rostellum following pollination did not prevent stigmatic closure in *Cymbidium*, but it does suppress anthocyanin formation in both column and labella [19]. In contrast, removal of the tip of the column (gynostemium) reduced the column swelling but not the stigmatic closure. Exposure of *Cymbidium* flowers to 10 µl of ethylene per litre for 24 h induced all the symptoms of post-pollination phenomenon. In addition, ethylene brought about colour changes in calli and wilting of perianth, but not straightening of gynostemium and stigmatic closure [18].

7.3 Postharvest Losses in Orchids

7.3.1 Nature of Losses

The condition of the flower at the time of the sale is a deciding factor in the marketing of orchids. Therefore, bruising, tearing of petal, discolouration, wilting or fading will render the flowers unsaleable. Furthermore, premature fading of flowers occurs when the flowers are pollinated, when the pollinia are disturbed or removed, or when the blossoms are exposed to ethylene.

Unfavourable light and temperature conditions, apart from delaying the flowering, reduce the quality of the flower.

Orchids also vary in their requirements for the pH of the potting mixture. Whereas the *Phalaenopsis* prefer a pH around 6.0, the *Cymbidium* requires a pH between 5.5 and 6.0 [2]. Any variation in pH is usually reflected in the number and quality of flowers. In the case of *Cattleya*, overmoist conditions during flowering affect the long-lasting quality of the bloom and increase the incidence of yellowing and rotting of flowers.

In periods of high humidity, particularly during fall and spring, when heat is not used at nights, the yellow pigmented areas in the throat of *Cattleya* often turn black after cutting. Fungal infection can also cause such spots in the throat. In contrast, in periods of hot and humid weather, an exudate appears on the lip of the orchids which causes discolouration and collapse of the tissue. The dark-coloured types are more vulnerable to this disorder. High humidity, particularly in industrial areas where smog is common, produces "dry sepal" in *Cattleya*. The varieties differ in their response to high humidity and smog.

Low temperatures (below 13.7 °C) cause chilling injury in *Cattleya, Cymbidium, Cypripedium,* and *Vanda*. The injury is characterized by the throat turning brown, the tip of the column turning black, and the development of water-soaked spots or areas on petals and sepals.

Some of the orchids, like *Cattleya*, produce deformed flowers at temperatures lower than 12.7 °C, which is quite often noticed during attempts to delay flowering.

Root rot, *Botrytis* rot, brown rot and mosaic are some of the more serious diseases of orchids. Root rot is very common with *Cattleya*, especially during cold and wet weather. The conditions favourable for the development of *Botrytis* are similar to those for root rot, but it becomes very severe when water drops are formed on the flowers. The ethylene liberated by the fungus results in the fading of the whole flower. Some of the viral diseases in *Cattleya* cause breakdown of flower colour, thus affecting the marketability of the affected flower.

Several insects, like aphids, thrips, scales, snails, slugs and ants, cause direct damage to the plant tissues, and stimulate ethylene production by the affected organ, thereby increasing the extent of damage.

7.4 Loss Reduction Biotechnology

Loss reduction biotechnologies in orchids involve the use of improved methods of harvesting, packing and transportation, use of low temperature, low pressure and controlled atmospheric storage, control of pests and diseases and the use of chemicals in order to inactivate the liberated ethylene.

7.4.1 Harvesting, Packaging and Transportation

The stage of harvesting varies with the type of orchid. *Cattleya* is harvested 3 to 5 days after the bud splits or dehisces. The weather conditions prevailing at the time of harvesting influence the exact stage of picking. During warm weather, the blossoms can be harvested at an earlier stage of development. Tagging the buds each day facilitates the cutting process.

Phalaenopsis can be cut when fully open. Individual flowers or the entire spikes are harvested at one time. *Cymbidium* is also cut in a similar way as *Phalaenopsis*. In the case of *Cymbidium*, it is very common to cut the whole spray and then separate the individual flowers at the time of grading.

The flowers must be placed in warm water (12.7 °C) as soon as possible after harvesting. *Cattleya* blooms develop spots on the petals if they are held at temperatures lower than 10 °C.

The flowers are shipped either as intact sprays or as individual flowers. When individual flowers are to be shipped, the flower stems are inserted through rubber caps into a tube containing water. Proper precautions are necessary to avoid any injury to flower stems or blooms. These tubes are taped or fixed to the bottom of the box and secured properly to avoid any injury due to displacement of tubes and/or flowers. It is also necessary to provide enough protection around individual flowers by using shredded paper so that they do not rub against either each other or rough surfaces in transit.

The small intact sprays are packed in corrugated cartons of varying sizes. The high relative humidity needed for preventing wilting of flowers can be maintained by providing a polythene lining to the sides and bottom of the carton, and folding it on the top before the carton is sealed for shipment.

7.4.2 Low Temperature Storage

Orchids are not adapted to low temperature storage. Most of the genera store best at temperatures between 4.8 and 7.8 °C, with the exception of few, like *Vanda*, which require higher temperatures (13.7 °C) [3]. Lower temperatures often produce chilling injury (darkening) on column and labellum. In extreme cases, the sepals and petals are also affected. Under normal conditions (4.8 to 7.8 °C), most orchids can be stored for 2 weeks. However, certain types of cultivars (certain types of *Cymbidium*) can be successfully stored at −0.6 °C without any chilling injury or loss in quality.

It is not advisable to store orchids for extended periods of time because most of the types last even longer or better when kept on the plant itself. This is particularly the case with certain varieties of *Paphiopedilum* and *Cymbidium*, which last approximately 6 to 7 weeks on the plant but only 2 to 3 weeks after they are harvested.

The flowers can be stored dry. However, pre-conditioning of flowers in 28.8 °C water and 4.8 °C air temperature for 24 h before dry storage is beneficial for successful storage [3]. Ethylene levels of only 0.002 ppm in the storage environment promote "dry sepal injury" and bleaching of sepals. Therefore, it is extremely important to inactivate the ethylene present in the storage atmosphere before introducing orchid flowers.

7.4.3 Controlled Atmospheric (CA) Storage

The possibility of using controlled atmospheric storage for delaying the fading of normal flowers in the presence of natural ethylene or ethylene-producing

Table 7.1. Effects of 2-day storage under 1.5% and 2.0% CO_2 on fading of *Vanda* flowers [9]

Initial CO_2 conc. (%)[a]	Percent ethylene generator (EG) and normal flowers (NF) faded[b] on day										
	0[c]			1	2	3	4	5	6	7	8[d]
	EG	NF	EG	NF	NF	NF	NF	NF	NF	NF	NF
0.03 (air)	100	100									
1.50	0	0	100	0	0	0	0	2.8	6.5	13.0	17.6
2.00	0	0	100	0	0	0	0	4.6	8.3	13.0	16.7

[a] 108 normal flowers and 9 ethylene generators per jar (6.3 l).
[b] Individual flowers 25 percent or more faded.
[c] Day removed from CO_2 storage.
[d] All remaining flowers wilted.

Table 7.2 Effects of 3-day storage under various concentrations of O_2 maintained with N_2 on fading of *Vanda* flowers [9]

Initial O_2 conc. (%)[a]	Percent degree of fading of ethylene generators (EG) and normal flowers (NF)											
	0[b]		1		2		3		4	5	6	7
	EG	NF	EG	NF	EG	NF	EG	NF	NF	NF	NF	NF
0.0	10[c]	0	50	0	90	0	100	12.5±2.76[c]	12.5±2.76	12.5±2.76	12.5±2.76	
1.0	50	0	70	0	90	0	100	0	12.5±2.99	12.5±2.99	37.5±7.08	
2.6	0	0	10	0	75	0	100	0	0	0	12.5±2.72	50.0±50.0[c]
3.2	25	12.5±2.91	87	52.75±7.16		52.75±7.16	100	55.0±5.92	55.0±5.92	66.0±4.00	87.5±0.72	90.0± 1.29
5.2	95	82.5±1.71	100	100	–	–	–	–	–	–	–	–
21.0 (air)	100	100	–	–	–	–	–	–	–	–	–	–[b]

[a] 10 normal flowers and one EG per jar (1.8 l).
[b] Day removed from O_2 storage.
[c] Flowers wilted or brown.

Table 7.3. Effects of 3-day storage under various pressures on fading of *Vanda* flowers [9]

Initial pressure (mmHg)[a]	Percentage of fading of ethylene generating (EG) and normal flowers (NF) on day										
	0[b]		1		2		3	4	5	6	7[c]
	EG	NF	EG	NF	EG	NF	NF	NF	NF	NF	NF
125	0[c]	0	70	0	100	0	0	0	0	0	5.1
185.5	25	12.5	95	47.5	100	47.5	55.0	55.0	67.5	77.5	90.0
760	100	100	–	–	–	–	–	–	–	–	–

[a] Ten normal flowers and one ethylene generating flowers per jar (1.79 l).
[b] Day removed from pressure storage.
[c] All remaining flowers wilted.

flowers was studied by Akamine and his co-workers [9]. A static exposure for 2 to 3 days to 1.5% to 2.0% CO_2, or 1.0% to 2.6% O_2 modified with nitrogen effectively delayed fading for several days in the open (Tables 7.1 and 7.2). The residual effect of CA storage was evident even after the flowers were removed from storage. CO_2 acted as a competitive inhibitor of ethylene and inactivated the effects of ethylene produced by other flowers [9].

Packaging of flowers in modified atmospheres created by using nitrogen or CO_2 is also effective in preventing fading of flowers in the presence of ethylene-generating materials. For shipping under modified atmosphere, the packaging materials must be impervious to gases and durable to withstand rough handling in transit.

7.4.4 Low-Pressure Storage

Reduced pressures (125 to 188.5 mmHg) are effective in delaying fading of orchid flowers. The effect of hypobaric atmosphere is given in Table 7.3. Reduction in the atmospheric pressure to 125 mmHg delayed the fading of normal flowers, as well as that of ethylene generators. The fading of flowers which were initially stored under 125 mmHg was only 8% after 7 days in the open. Reduced pressure storage has been recommended for holding the flowers in storage before shipping.

7.4.5 Use of Ethylene Scrubbers

Chemicals like potassium permanganate ($KMnO_4$) have been successfully used to inactivate ethylene and to delay ripening in banana. In the case of *Vanda* orchids, Akamine and Goo [8] demonstrated the feasibility of using $KMnO_4$, Purafil and brominated activated charcoal in preventing or delaying fading of blossoms. $KMnO_4$ was effective both as a solution (free) or a solution absorbed on inert material (gauze, cardboard, cotton, paper towelling, cotton cloth, facial tissue, perlite and vermiculite) (Table 7.4). Direct contact

Table 7.4. Effects of potassium permanganate on ethylene (C_2H_4) production and fading on *Vanda* orchids [8]

$KMnO_4$ conc. (%)[a]	Days for normal flowers to fade[b]	C_2H_4 production (ppm) on day					
		1	2	3	4–10	11	12
2.0	11.8	0	0	0	0	2.33	0.78
1.0	2.7	0	7.77	5.44			
0.5	2.5	0	6.22	6.99			
0.0	1.8	6.22	20.59	52.84			

[a] Fifteen normal flowers with one ethylene-generating flower sealed in 6.3-l jar.
[b] Ethylene-producing flower faded in 1 day in all treatments.

Table 7.5. Comparative effectiveness of Purafil and 20% brominated activated charcoal for controlling fading of *Vanda* flowers [8]

Treatment[a]	Percentage normal flowers faded on day[b]				
	1	2	3–8	9	10–19[c]
5 g Purafil	0	0	0	0	0
5 g Brominated activated charcoal	0	0	9.1	100	
Control	0	100			

[a] Eleven normal flowers with one ethylene-generating flower sealed in 6.3-l jar.
[b] Ethylene-generating flowers faded in 1 day in all treatments.
[c] Terminated on day 19 because of decay from ethylene-producing flowers contaminated normal flowers.

Table 7.6. Effects of Purafil on fading of *Vanda* flowers [8]

Purafil (g)	Jar vol. (l)[a]	Percentage of normal flowers faded on day[b]			
		1	2	3–15	16–20[c]
0	1.8	0	100		
5	1.8	0	0	0	0
10	1.8	0	0	0	0
25	1.8	0	0	0	0
0	6.3	0	100		
5	6.3	0	0	0	8.3
10	6.3	0	0	0	0
25	6.3	0	0	0	8.3

[a] Twelve normal flowers with one ethylene-generating flower was sealed in each jar.
[b] Ethylene-generating flowers faded in one day in all treatments.
[c] Terminated on day 20 when decay from ethylene-generating flowers contaminated normal flowers.

of the flowers with $KMnO_4$ causes injury to petals. $KMnO_4$ impregnated in perlite or vermiculite was also very effective in delaying fading of *Vanda* flowers. A commercial preparation, Purafil (activated alumina pellets impregnated with $KMnO_4$), was easier to use than perlite or vermiculite and was equally effective in controlling the fading or normal flowers in the presence of C_2H_4-generating flowers in package (Table 7.5). It is also easy to determine the effectiveness of Purafil, as the colour of pellets turns darker (dark purple) with decreasing efficiency. Direct contact of flowers with $KMnO_4$ can be easily avoided by placing Purafil pellets in a non-hygroscopic pouch. The amount of material to be used depends upon the size of the package and the number of normal and ethylene-generating flowers in a container or box (Table 7.6).

7.4.6 Control of Pests and Diseases

Maintenance of proper hygiene both in growing and postharvest environment is essential for producing quality flowers. Postharvest diseases are difficult to control once they become established. Therefore, a timely spraying schedule is of the utmost importance.

References

1. Bailey LH (1963) The Standard Encyclopedia of Horticulture. McMillan, New York USA
2. Laurie A, Kiplinger DC, Nelson KS (1968) Commercial Flower Forcing. McGraw-Hill New York
3. Staby GL, Robertson JL, Kiplinger DC, Conover CA (1978) Chain of life. Ohio Flor Assoc, The Ohio State Univ, Columbus
4. Anon (1982) Floriculture crops production, area, sales. Statistical Rep Serv USDA Washington DC
5. Rotor GB Jr (1952) Day length and temperature in relation to growth and flowering of orchids. Cornell Agric Exp Sta Bull 885
6. Sheehan TS, Murashige T, Kamemoto H (1965) Photoperiodism effects on growth and flowering of *Cattleya* and *Dendrobium* orchids. Am Orchid Soc Bull 34:228
7. Akamine EK (1963) Ethylene production in fading *Vanda* orchid blossoms. Science 140: 1217
8. Akamine EK, Goo T (1981) Controlling premature fading in *Vanda* Miss Joaquim flowers with potassium permanganate. Univ of Hawaii, College of Tropical Agric and Human Resources, Res Rep No 8
9. Akamine EK, Goot T (1981) Effects of static controlled atmosphere and reduced pressure storage on facing of *Vanda* Miss Joaquim flowers. Hawaii Inst Trop Agric and Human Resources, Honolulu No 8
10. Akamine EK (1976) Postharvest handling of tropical ornamental cut crops in Hawaii. HortScience 11:125
11. Lindner RC (1946) Studies on packaging and storage of *Vanda* (Joaquim) flowers. Hawaii Agric Exp Sta Progress Notes 49
12. Davidson OW (1949) Effects of ethylene on orchid flowers. Proc Am Soc Hortic Sci 53:440
13. Fischer Jr CW (1950) Presence of ethylene gas poses problems in cut flower storage. Flor Rev 106(3814):31
14. Sheehan TJ (1954) Respiration of cut flowers of *Cattleya mossiae*. Am Orchid Soc Bull 23:241
15. Arditti J (1969) Postpollination phenomena in orchid flowers. Aust Orchid Rev 34:155

16. Arditti J, Flick BH, Jeffrey D (1971) Postpollination phenomena in orchid flowers – II. Induction of symptoms by abscisic acid and its interactions with auxin, gibberellic acid and kinetin, New Phytol 70:333
17. Arditti J, Hogan NM, Chadwick AV (1973) Postpollination phenomena in orchid flowers – III. Effects of ethylene. Am J Bot 60:883
18. Arditti J, Flick BH (1974) Postpollination phenomenon in orchid flowers – V. Participation by the rostellum and gynostemium tip. Am J Bot 61:643
19. Arditti J, Flick BH (1976) Postpollination phenomenon in orchid flowers – VI. Excised floral segments of *Cymbidium*. Am J Bot 63:201
20. Arditti J, Knauff RL (1969) The effects of auxin actinomycin D, methionine and puromycin on post pollination behaviour in *Cymbidium* (orchidaceae) flowers. Am J Bot 56:620
21. Arditti J, Harrison CR (1979) Postpollination phenomena in orchid flowers. VIII. Water and dry weight relations. Bot Gaz 140 (2):133
22. Arditti J (1971) The place of the orchid pollen poison and pollen hormone in the history of plant hormones. Amer J Bot 58:480

Chapter 8 Gladiolus

8.1 Introduction

The name *Gladiolus* was suggested by Pliny to describe the shape of the leaf, which resembles that of a sword (Latin gladus-sword) [1]. It is grown from corms both for cut flower and garden decoration. As a cut flower, *Gladiolus* is one of the most important grown in Europe and the USA. The flowers vary widely in colour, with attractive shades like crimson, pink, salmon, scarlet, purple, cherry red, apricot, cream, white or a combination of two or more shades. In form, they are either single or double, with or without ruffled petals and pigmentation in the centre.

The *Gladiolus* flowers are suitable for large and lavish bouquets. These flowers offer certain definite advantages for flower arrangements [2]:

1. The spikes form a strong natural line in floral arrangement.
2. Flowers open acropetally, forming a natural progression of different stages in flower opening.
3. The stems are solid and can be used for variety of uses.
4. The leaves can also be used in floral arrangements.

The demand for cut *Gladiolus* is consistent throughout the year. Furthermore, the bulb production constitutes a sizeable proportion of trade in several western countries.

8.1.1 Postharvest Quality Requirements

Spike length, number, size, shape, and colour of floret, weight and length of flower stalk are some of the criteria used for determining the quality of cut flowers in *Gladiolus*. Out of these considerations, length of spike is the most commonly employed for grading *Gladiolus*, according to which the spikes are graded as fancy, special, standard, and utility [2]. Other postharvest quality considerations that need to be emphasized are [2]:

1. Florets should be uniformly spaced along the stem in proportion to the length of the stem. The inflorescence should have a balance between developed (opened), partially opened and immature buds at the time of sale.
2. Florets should be turgid, free from injury and should face in one direction only. The flower opening should be uniform.
3. Flower stems should be strong, straight and free from side shoots, and the length according to cultivar.

4. Foliage should be of proper length and free from damage.
5. Flower shape and size should be representative of the cultivar. They should open slowly.
6. Petals should be free from any discolouration, drooping or wilting.

8.1.2 Cultivars of Commercial Importance

Gladiolus cultivars are classified by a three digit number [3] (Table 8.1). The cultivars vary in their potential-lasting qualities. Certain cultivars, like White Friendship, Wild Rose, Spic and Span and Traveller have excellent postharvest qualities, while others like Tequendana, Hopman's Glory and Orange Red are at the other extremity (short-lived) [4]. Soil fluoride level also plays an important role in flower quality. *Gladiolus* cultivars show differential response to soil types and atmospheric fluoride contents [4].

Out of the 20 cultivars tested by Wilfret [3], only seven cultivars viz., Beverly Ann, Friendship, Spic and Span, Hopman's Glory, Traveller, Valeria and White Friendship were found suitable under Florida conditions. Wild Rose was found to have the best-quality flowers (a quality index of 6.1), while Orange Gold had the lowest flower quality index (0.6). Other cultivars commercially grown are T-210, Pink Purade, Intrepid, Red Majesty, Jester, Jacksonville

Table 8.1. Size and colour classification of *Gladiolus* cultivars developed by the North American *Gladiolus* Council [3]

Class [a]	Designation	Floret size (cm)
100	Miniature	6.4 and below
200	Small to miniature	6.4 – 8.9
300	Decorative	8.9 – 11.4
400	Standard or large	11.4 – 14.0
500	Giant	14.0 and above

Colour [b]	Pale	Light	Medium	Deep	Other
White	00				
Green		02	04	16	
Yellow	10	12	14	16	
Orange	20	22	24	26	
Salmon	30	32	34	36	
Pink	40	42	44	46	
Red	50	52	54	56	58 Black red
Rose	60	62	64	66	68 Black rose
Lavender	70	72	74	76	78 Purple
Violet	80	82	84	86	
Smokies		92	94	96	
Tan	90				98 Brown

[a] The first digit indicates the floret size in the five classes.
[b] The first two digits in the classification indicate floret colour and shade. An odd last digit indicates a conspicuous mark or blotch.

Gold, Goldfield, Flamingo Peter Pears, Impala, Spring Song, King Ramson, Truelove, White Prosperity, Morning Bride, Lemon Lime, Golden Harvest and Golden Scepter [3].

8.1.3 Production

Gladiolus is a leading bulbous cut flower in Poland, while it ranks second in the Netherlands. Other centres of production are Germany, the USA, the UK, Israel, France and Italy. In the USA, *Gladiolus* is grown in the open in California and Florida. However, almost all of the gladioli produced in European countries are from plastic or greenhouses [3].

8.2 Physiology and Biochemistry

8.2.1 Growth and Development

Unlike other bulbous crops, flower initiation in *Gladiolus* occurs in the early stages of plant growth in the field [5, 7]. Flower induction in this case is referred to as autonomous, in which the initiation is always preceded by the production of a definite number of leaves (8 to 10). Subsequent development of floral buds greatly depends on the prevailing environmental conditions. Light, temperature and soil water levels are critical for normal flower bud development [6, 8–16]. Of these, day length and light intensity are considered to be most important for *Gladiolus* flowering [13, 14, 26].

These environmental factors have both qualitative and quantitative effects on flowering in *Gladiolus* [16]. Floret initiation begins when the plants are 30 to 40 mm tall and continues until the expansion of the 7th leaf [15]. Sub-optimal growing conditions during this period result in blasting of individual florets or of whole inflorescence [12]. The extent of damage is governed by the

Table 8.2. Effects of coold nights on flower quality in Sans Souci *Gladiolus* [a] [15]

Developmental stages	Flowering (%)	Stalk length (cm)	Spike length (cm)	No. of florets per spike
Planting	25	–	–	–
Sheath – 2-leaf	100	131.6 ± 3.5	42.8 ± 2.2	11.6 ± 0.5
1 – 3-leaf	100	133.6 ± 3.7	42.4 ± 1.2	13.6 ± 0.2
2 – 4-leaf	100	134.4 ± 2.6	48.4 ± 1.6	13.2 ± 0.7
3 – 5-leaf	100	128.6 ± 3.8	45.6 ± 1.5	12.4 ± 0.5
4 – 6-leaf	100	127.6 ± 3.3	47.4 ± 0.8	14.0 ± 0.8
5 – 7-leaf	100	128.0 ± 2.0	52.2 ± 2.2	13.6 ± 1.0
6 – spike emergence	100	115.6 ± 1.6	49.4 ± 2.2	12.2 ± 0.9
7 – anthesis	80	90.6 ± 6.7	27.2 ± 5.3	5.5 ± 1.8
Control	96	111.0 ± 6.0	39.0 ± 3.7	11.6 ± 1.1

[a] Plants exposed to 1 °C for 16 h continuously for 23 nights during summer.

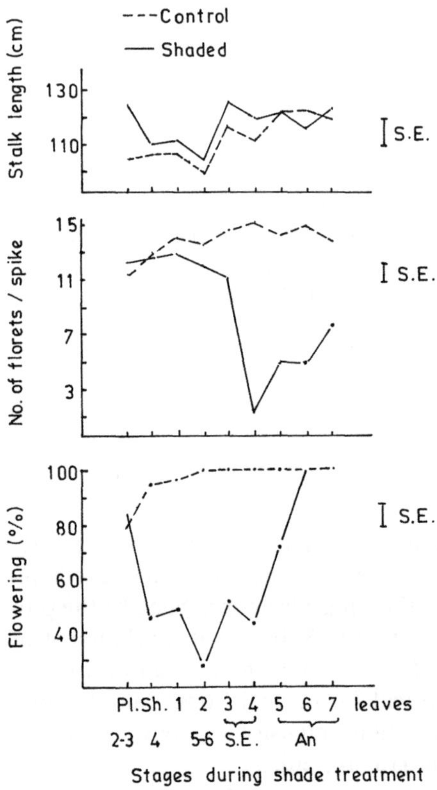

Fig. 8.1. Effects of shading (95%) applied for 26 days starting at different stages on flowering of Sans Souci plants [13]

severity and duration of such unfavourable conditions. The sensitivity of plants to these conditions depends upon the stage of development at the time of exposure. In the early stages of growth (from planting to sheath leaf stage), *Gladiolus* plants were found to be very sensitive to soil moisture stress and high temperatures [15]. The number of florets per spike and percentage of flowering is affected if the conditions are not favourable (Table 8.2). Low soil temperatures at this stage affect flower quality [14]. The plants have the highest rate of cell division and leaf primordia initiation at this stage. In certain cases, the apices are transformed in the prefloral stages and hence the adverse environmental conditions result in flower bud blasting [12, 14]. In contrast, plant in the later stage of development (from sheath leaf to 1st leaf stage), when the apices change from the prefloral to floral stage, are comparatively more tolerant to these unfavourable conditions [12].

The sensitivity of plants to sub-optimal environmental conditions changes considerably with age. *Gladiolus* in the one- to four-leaf stage are affected most by unfavourable light intensity and duration (Figs. 8.1 and 8.2) [14, 15]. In contrast, sensitivity to temperature and water stress increases with age from the first leaf stage, reaching maximum at the four- to six-leaf stages [15, 16]. Low light intensity, together with low temperatures at this stage, has most

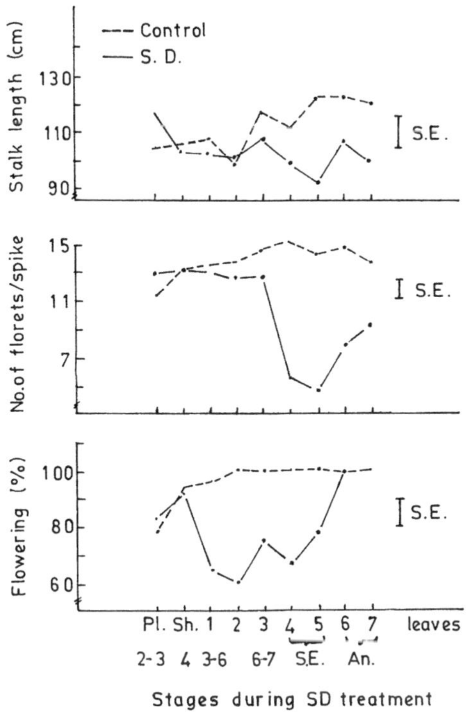

Fig. 8.2. Effects of photoperiod (26 short days) applied at different stages of development on flowering of Sans Souci gladioli [14]

disastrous effects on flower quality, as it affects the initiation of florets and stalk enlargement [12]. Plants that escaped damage at this stage generally bear normal quality florets. The damaging effects of unfavourable climatic conditions continue until spike emergence and, under severe conditions, the whole spike may be damaged [15]. At the final stages of flowering (spike emergence to anthesis), the plants are more sensitive to high temperatures than to low light. Water deficit affects the flower quality at the time of spike emergence only. The reduction in the plant's sensitivity to water stress at the final stages of flowering has been correlated with the translocation of photosynthates from leaves to flower stalks and to the decreased dependence on soil moisture [16].

Plants vary in their response to these conditions with the season. In general, they are more sensitive to low light and water stress during winter than in the warmer months [16]. Long photoperiods (beyond 12 h) at the one- to two-leaf stage increase the number of florets per spike, the length of spike, and flower stalk, but delay flowering. In contrast, short photoperiods at this stage accelerate flowering, but restrict accumulation of photosynthates by reducing the growing period. The total yield of marketable spikes is also adversely affected at short photoperiods, especially during winter.

The corms are dormant at the time of digging. The length of the dormant period varies with the cultivar, size of corm and environmental factors, like soil temperature at the time of digging [9]. In general, larger-sized corms have shorter rest periods than smaller ones and cormels. The corms of certain florist

Gladiolus cultivars remain dormant for 12 to 18 months at about 26.4 °C and the length of dormancy is correlated closely with increase in temperature during storage. Dormancy can be broken by placing the corms at 4° to 5 °C for 6 weeks. Certain ethylene-releasing chemicals are also being used to break the dormancy [3].

8.3 Postharvest Losses in Cut Gladiolus

Postharvest quality losses in *Gladiolus* are related to the use of unhealthy planting material, sub-optimal growing conditions, pre- and postharvest pests and diseases, physiological disorders and faulty harvesting and handling techniques. The cut flowers are discarded when the bottom five florets are wilted, which results in wastage of the rest of the florets that are still good or yet to open. Under such conditions, the postharvest losses will be well over the standard 20%.

Causes of Losses. The majority of the postharvest losses in quality as well as quantity in cut *Gladiolus* are caused by one or more of the following factors:

1. Cultural.
2. Mechanical.
3. Physiological disorders.
4. Pests and diseases.

These factors are interrelated, and may increase the incidence of losses due to other factors. For example, sub-optimal cultural conditions may reduce the quality of flowers, as well as decrease their storageability and increase susceptibility to postharvest diseases and pests.

1. Cultural Factors

Photoperiod, solar radiant flux, temperature and soil moisture content play a major role in determining the quality of the flower. The reduction in solar radiant energy, especially during the early stages of growth, increases bud "blasting" or affects the initiation of florets. The plants in the sheath-leaf to the one-leaf stage are also sensitive to high temperatures and water stress. In both cases the percentage of blind shoot is increased. Photoperiods longer than 12 h given at the one- to two-leaf stages accelerate flowering, but decrease flower quality. The shorter photoperiods at this stage delay flowering and increase percentage of blind shoots [16] (Table 8.3). However, the varieties vary in their response to photoperiod.

Low night temperature during plant development also increases the incidence of flower blasting. Spike length and number of florets per spike are drastically reduced if the plants are exposed to low night temperatures at the seventh leaf stage to spike emergence. In contrast, low temperatures immediately after planting resulted in a maximum number of blind shoots [15].

Table 8.3. Effects of shortening natural day length on the flowering and flower quality in cut gladiolus [15]

Cultivar	Day length (h)	Flowering (%)	Number of florets per spike	Number of secondary inflorescence per spike
Dry Fleming	11 – 13 (Natural)	82	14.6	0.7
(Expt. No. 1)	8	71	10.0	0.0
	6	57	10.0	0.2
	4	22	10.0	0.0
Sans Souci	$13\frac{1}{2} - 14\frac{1}{2}$ (Natural)	91	23.7	0.2
(Expt. No. 2)	6	23	44.5	0.0
Spic and Span	$13\frac{1}{2} - 14\frac{1}{2}$ (Natural)	97	17.4	0.9
(Expt. No. 2)	6	93	14.2	0.0

Table 8.4. Effects of moisture stress on flowering and flower quality in Sans Souci *Gladiolus* [15]

Stage of development during treatment	Flowering (%)	Stalk length (cm)	Number of florets per spike
Planting to sheath leaf	0	–	–
Sheat to 2-leaf	100	105	11.2
1 – 2.5-leaf	60	115	13.0
2 – 3.5-leaf	20	80	6.0
3 – 4.5-leaf	25	80	5.0
4 – 5.5-leaf	0	–	–
5 – 6.5-leaf	0	–	–
6 – 7.0-leaf	20	93	5.0
7 – Spike emergence	40	93	6.0
Control	100	110	11.2

In general, moisture stress is detrimental to plant growth. Moisture stress results from excessive transpiration of water in comparison to absorption. Occurrence of water stress is related to irrigation frequency, which reduces cell enlargement almost instantly. The extent of damage depends on the duration, severity and stage of development at the time of occurrence of moisture stress. In the long run, it reduces the rate of photosynthesis and protein synthesis, and in severe cases it inhibits cell division. In *Gladiolus*, plants that are in the early stages of development when flower initiation occurs, or those in later stages when flower petals undergo rapid expansion are most affected by water stress. The moisture stress, if very severe, even for a short time, can completely wipe out the whole crop. The sensitivity of Sans Souci *Gladiolus* to water stress at defferent stages of development is given in Table 8.4.

Nitrogen dificiency reduces number of florets per stems, while potassium deficiency results in lower floral bud count, reduction in stem length, and causes delay in flowering [35]. Calcium deficiency causes cracking of the spike generally below the second or third floret and, in severe cases causes bud blasting [35]. Similarly, boron deficiency causes reduction in spike length.

2. Mechanical Factors

Mechanical loss results mainly from bruising, crushing or tearing of petals, breaking of florets or spikes. These injuries, apart from causing direct losses due to wastage, accelerate ethylene synthesis in the injured tissue and in severe cases may render the whole lot unsalable.

3. Physiological Disorders

"Blindness", "topple" and "negative geotropic curvature" are some of the more common physiological disorders in cut *Gladiolus*. Apart from these disorders, postharvest losses in quality can be traced back to mineral deficiencies or toxicities, physiological wilting, or vascular blockage in cut stems etc. Apart from the direct effects on the flower quality, these physiological disorders can render the plant tissues susceptible to damage caused by other factors.

"Blindness" in cut *Gladiolus* is caused by unfavourable growing conditions during flower bud initiation and development [16]. *Gladiolus* "topple" is a postharvest disorder characterized by breaking of spikes as the florets open in vases. This disorder is closely correlated with calcium content in the stalk tissue. The spikes from plants grown in calcium-deficient soil have a higher incidence of "topple" [4].

Negative geotropism is another postharvest disorder that occurs during storage or shipping. The spikes curve upward if they are kept horizontally. Differential redistribution of auxins has been associated with this phenomenon. The bending in most cases is irreversible and renders the spike unfit for bouquet and floral arrangement. However, negative geotropic curvature has no effect on the lasting qualities of the flower [4].

4. Pests and Diseases

Cutworms, wireworms, armyworms, thrips, red spider mites, leaf hoppers and nematodes are some of the insect pests that are quite common in *Gladiolus* ranges or fields. These insects cause direct damage to plant tissues by chewing the foliage of sucking the juice. They also stimulate ethylene biosynthesis in the host tissues.

Several diseases are troublesome in *Gladiolus. Curvularia* causes rotting of infected tissues of leaves, stems, florets or corms. In severe cases, the infected florets fail to open normally. *Botrytis* is another fungus that attacks leaves, stems, and flowers [18] (Fig. 8.3). It is a disastrous fungus because of its ability to grow even at $-2\,°C$ [4]. High relative humidity coupled with warm conditions favour *Botrytis* growth. The infected tissues become covered with grey mould, which stimulates ethylene production in host tissues by up to four to five times the level in normal healthy tissue [14].

Stemphylium infection has been found to reduce flower size. The infected tissues turn bright yellow, and finally become transluscent on both sides [18]. *Fusarium* is by far the most destructive fungus. The symptoms of *Fusarium* infection vary from simple stunting to yellow leaves, malformed or restricted leaf growth, and weak flower stalks with small florets [18]. The petals in the infected florets may be deformed or malformed, with prominent veins. Since the fungus is soil-borne and transmitted through corms, the material must be tested for Fusarium infection before planting [18].

8.3.1 Cultivars

Cultivars vary considerably in respect to their potential-lasting qualities. The short-lived varieties like Hopman's Glory and Orange Gold are genetically inferior to others which last longer [4]. These cultivars also differ in their sensitivity to atmospheric fluoride levels. Therefore, selection of proper cultivars for growing is an important consideration in producing quality flowers.

Fig. 8.3. *Botrytis* flower blight in *Gladiolus* [18]

8.4 Loss Reduction Biotechnology

Qualitative and quantitative postharvest losses can be reduced by adopting newer biotechnologies, such as harvesting at proper stages, use of proper packing methods, and improved storage techniques, such as use of low temperature and modified atmosphere storages, use of floral preservatives and control of pests and diseases [4, 19–25].

8.4.1 Cultural Practices

The importance of optimum growing conditions for producing quality flowers has already been explained earlier. Growing conditions, as far as possible, should be optimum in order to produce quality flowers that can last longer. Proper crop rotation with non-Iridaceae species is beneficial for controlling some of soil-borne diseases. Selection of resistant cultivars is also an important step in the production of quality flowers.

8.4.2 Harvesting, Grading and Packaging

The stage of harvesting depends upon the prevailing temperature and the market outlets. Most growers harvest their crop when the bottom two florets show colour. While in cold weather the spikes can be cut at a later stage of plant development, in warm weather they will have longer postharvest life if they are cut at an early stage of development and open in in vitro conditions. Growers with facilities for bud opening harvest their crop when the buds swell and petals begin to show colour [4]. Flowers to be shipped over a long distance are cut in the tight bud stage, as this technique has several advantages.

Flowers are graded according to established standards. Several growers in the USA have developed their own grading system and employ it for their crop (Table 8.5). The spikes after grading are bunched in units of 10 and held together by rubber bands. For local markets flowers are shipped in upright position in boxes of hampers, but for long-distance markets 10-spike bunches are packed without water in hampers made of fibre-board or wood. Usually 15 to 24 bunches are packed in each hamper.

Table 8.5. Grades of cut flowers used by Florida commercial *Gladiolus* growers [3]

Grade	Spike length (cm)	Number of florets per spike (minimum)
Fancy	107 and above	16
Special	96–107	14
Standard	81–96	12
Utility	81 or below	10

8.4.3 Low Temperature Storage

The optimum temperature for long-term storage of cut *Gladiolus* is 1.7 to 4.4 °C [4]. Temperatures below 1.0 °C usually result in chilling injury. Most of the florist *Gladiolus* cultivars are not suitable for long-term storage, as the florets on removal from storage (14 days) usually do not open normally [23]. Under ideal conditions, cut gladioli can be stored for 5 to 7 days without any losses in lasting quality [4].

Preconditioning of flowers for 16 to 20 h in a solution containing 20% sucrose and 200 ppm Physan-20 improved the storageability and post-storage life of cut flowers [20]. The best temperature for pulsing is approximately 20 °C. Other preservatives recommended for the purpose are silver nitrate, silver thiosulfate and quarternary ammonium compounds [19–22, 27–34].

Temperature during long-distance transportation plays an important role in the subsequent performance of cut gladioli under indoor environment. The room temperature vase life of cut flowers of *Gladiolus* after 1 to 2 days transport was reduced from 6 to 8 days to 4 to 6 days by raising the temperature during shipping from 1.7–4.4 to 10.0 °C [18]. The opening of florets on arrival was also affected by increasing the storage temperature.

8.4.4 Controlled and Modified Atmospheric Storage

Floret opening and subsequent vase life can be improved by storing the flowers in a controlled environment [23]. The rate of respiration in CA-stored (5% CO_2 and 2% O_2) flowers was 20% to 30% lower than those held in normal air [23]. The floret opening, as well as the total number of florets opened, increased in CA-stored spikes.

Dry storage of flowers in polyethylene bags without lime was found to be the best for 6 to 8 days holding [23]. The polyethylene film prevented water loss and maintained high relative humidity. The flowers were injured in modified atmospheres created by increasing nitrogen concentration [23].

8.4.5 Sub-Atmospheric Pressure Storage

Research on sub-atmospheric pressure storage (hypobaric) of storage of cut *Gladiolus* is very scanty and inconclusive. According to the available data, the flowers can be stored over 30 days under sub-atmospheric conditions. The cost involved in this method does not warrant the use of hypobaric storage in the case of *Gladiolus*.

8.4.6 Use of Floral Preservative

Use of floral preservative is the most economical and practicable method for extending the postharvest life of *Gladiolus* cut flowers. The best-known preservative solution contains 600 ppm of 8-HOC and 4% sucrose [4]. The time elapsing between cutting and placing them in the preservative solution determines the effectiveness of the chemical in delaying senescence. Pulsing of flower stems in sucrose with other preservatives before storage or shipping improves floret opening [27–34].

8.4.7 Control of Pests and Diseases

An integrated pest control is necessary to maintain adequate hygiene. The pest control schedule should include both fungistates and fungicides which are effective in preventing the diseases from occurring. Furthermore, the planting materials must be tested before introducing them or planting.

References

1. Bailey LH (1963) The Standard Encyclopedia of Horticulture. McMillan, New York USA
2. McDaniel GL (1982) Ornamental Horticulture. Reston, Westport USA
3. Wilfret GJ (1980) Gladiolus. In: Larson RA (ed) Introduction to Floriculture. Academic Press, London
4. Staby GL, Robertson JL, Kiplinger DC, Canover CA (1978) Chain of Life. Ohio Flor Assoc, pp 71
5. Rees AR (1972) Growth of bulbs. Academic Press, London
6. Halevy AH (1972) Water stress and timing of irrigation. HortScience, 7:113
7. Pfeiffer NE (1931) A morphological study of *Gladiolus*, Contrib Boyce Thompson Inst 3:173
8. Halevy AH (1962) Irrigation experiment of *Gladiolus*. In: Vasturo PV (ed) Gladiolus. Maxifed p 129
9. Apte SS (1960) Study of some factors affecting rest and production of spikes, corms, and cormels in *Gladiolus*. M Sc Thesis Univ Poona
10. Kosugi K (1962) Studies on the production and flowering in *Gladiolus*. Mem Fac Agric Kagawa Univ 11:1
11. Shilo R, Halevy AH (1963) The effects of low temperature on flowering of gladioli. Isr J Agril Res 13:141
12. Shilo R, Halevy AH (1976a) Inflorescence development of flowering and blasted gladiolus plants in relation to development of other plant parts. Sci Hortic (Amst) 4:79
13. Shilo R, Halevey AH (1976b) The effects of various environmental factors on flowering of gladiolus – I light intensity. Sci Hortic (Amst) 4:131
14. Shilo R, Halevy AH (1976c) The effects of various environmental factors on flowering of gladiolus – II Length of the day. Sci Hortic (Amst) 4:139
15. Shilo R, Halevy AH (1976d) The effects of various environmental factors on flowering – III temperature and moisture. Sci Hortic (Amst) 4:147
16. Shilo R, Halevy AH (1976e) The effects of various environmental factors on flowering gladiolus – IV Interaction of environmental factors – general discussion. Sci Hortic (Amst) 4:157
17. Tonecki J (1980) Effect of growth regulators on shoot apex differentiation and changes in sugars and free amino acids in gladiolus (*Gladiolus hortorum* cv. Acca Laurentia). Acta Hortic 109:347
18. McKay ME, Hughes I (1982) Growing gladioli. Queensland Agric J 108:127
19. Mor Y, Hardenburg RE, Kofranek AM, Reid MS (1981) Effect of silver thiosulfate pretreatment on vase life of cut standard carnations, spray carnations, and gladiolus after a transcontinental truck shipment. HortScience 16(6):766

20. Kofranek AM, Halevy AH (1976) Sucrose pulsing of gladiolus stems before storage to increase spike quality. Hort Science 11(6):572
21. Halevy AH, Mayak S (1981) Senescence and postharvest physiology of cut flowers. Hort Rev Vol 3
22. Farhoomand MB, Kofranek AM, Mor Y, Reid MS, Awad ARE (1980) Pulsing of *Gladiolus hybrida* Captain Bush with silver or quarternary ammonium compounds before low temperature storage. Acta Hortic 109:253
23. Hardenburg RE, Uota M, Parson CS (1969) Refrigeration and modified atmosphere for improving keeping quality of cut flowers. Proc 12th Int Congr Refrig 3:339
24. Fischer CW (1953) Long term holding of cut flowers. Proc Am Soc Hortic Sci 61:585
25. Waters WE (1967) Well water has strong effect on cut chrysanthemum, gladiolus and roses. Sunshine State Agric Res Rep 12:4
26. Groen NPA, Rotten LAJM van de (1980) Energy saving in flower production of gladiolus forced in glasshouses. Acta Hortic 109:117
27. Bravdo B, Mayak S, Gravieli Y (1974) Sucrose and water uptake from concentrated sucrose solutions by gladiolus shoots and the effects of these treatments on floret life. Can J Bot 52:1271
28. Mayak S, Bravdo B, Gvilli A, Halevy AH (1973) Improvement of opening of gladioli flowers by pretreatment with high sugar concentrations. Sci Hortic (Amst) 1:357
29. Kofranek AM, Paul JL (1974) The value of impregnating cut stems with high concentrations of silver nitrate. Acta Hortic 41:199
30. Marousky FJ (1968) Influence of 8-hydroxyquinoline citrate and sucrose on vase life and quality of cut gladiolus. Proc Fla Sta Hortic Soc 81:415
31. Marousky FJ (1969) Conditioning gladiolus spikes to maintenance of fresh weight with pretreatments of 8-hydroxy quinoline citrate plus sucrose. Proc Fla Sta Hortic Soc 82:411
32. Marousky FJ (1971) Effects of temperature, container venting and spike wrap during simulated shipping and use of floral preservative on subsequent floret opening and quality of gladiolus. Proc Trop Reg Am Soc Hortic Sci 15:216
33. Marousky FJ, Woltz SS (1971) Effect of fluoride and a floral preservative on quality of cut gladiolus. Proc Fla Sta Hortic Soc 84:375
34. Waters WE (1966) The influence of post-harvest handling techniques on vase life of gladiolus flowers. Proc Fla Sta Hortic Soc 79:452
35. Woltz SS (1955) Effect of differential supplies of nitrogen, potassium and calcium on quality and yield of gladiolus flowers and corms. Proc Am Hortic Sci 65:427

Chapter 9 Other Cut Flowers

9.1 Gerbera

9.1.1 Introduction

Gerbera (*Gerbera jamesonii*, Hook.), named in honour of Traug Gerber, is commonly known as Transvaal Daisy, or Barberton Daisy. It is a dwarf perennial herbaceous plant growing in clumps with solitary flower heads on long slender stems which grow well above the foliage. The flowers come in a wide range of colours including yellow, white, orange, pink, crimson and purple. Gerberas are good for beds, borders, rock gardens, pot culture and cut flowers. The flowers suit very well in different floral arrangements. Gerberas can be successfully grown both on plains and in the hills. In western countries they are usually grown in greenhouses, but in mild weather are even grown outdoors on the south-west side. Cultivars of commercial importance are Maroon, the Clementine series, Xenia, Flamengo, Delphi, Apple Blossom and Uranus [1].

Photoperiod and temperature during vegetative growth have tremendous effects on the production of lateral shoots and flowers. Low temperatures (17 °C night) and 8 h photoperiod have positive correlation with lateral shoot and flower production. Most productive plants have been shown to contain two to five lateral shoots during the first year and a leaf-to-flower ratio of 2 : 4 [2, 4]. The number of leaves per flower range from 1 – 11. Flowering was reduced when the leaf-to-flower ratio was increased.

9.1.2 Senescence of Gerbera Cut Flowers

Senescence of cut *Gerbera* inflorescence is caused by continuous changes in water content and ion leakage in the petal tissues [5 – 10]. The water content of petals shows an initial increase a few days after cutting from the parent plant and then a very rapid decrease coinciding with wilting and death of the inflorescence [6]. The decrease in dry matter of petals in the first few days in the vase has been correlated with increase in water content [6]. The rapid decline coincides with the reduction in the permeability characteristics of the membrane in petal cells and an increase in ion leakage which causes a reduction in the water-holding capacity of the petals [6]. A similar trend was not observed in intact flowers. Lowering of pressure potential (ψ press) in excised petals stimulated the increase in ion leakage, while increasing the ψ press by placing the stems in potassium nitrate or 20% sucrose solution (24 h) delayed it [10].

Different cultivars show different degrees of response to cutting, which are reflected by differences in the moment at which the increase in ion leakage in the petals occurs. In the longer-lasting cultivar, Mini Wit, the increase in ion leakage started 12 days later than in short-lived cultivars like Wagningen Rood and Citronella [6]. The petals of cv. Mini Wit showed have the highest ψ press of all the three cultivars [8].

The season of flowering also influences the keeping quality of the flower. Inflorescences harvested during summer (April) contained less water and had higher ψpress than those harvested in the months of September and October. Summer flowers also had better keeping qualities than the autumn flowers [10]. Meeteran [10] suggested a strong relationship between ψ press (rigidity) of petals on the day of harvesting and number of days after cutting until the increase in ion leakage occurs.

9.1.3 Postharvest Losses in Cut Gerberas

The postharvest losses in cut *Gerbera* inflorescence result from genetic, cultural, mechanical and physiological disorders. Rough handling during harvesting, packing and/or transportation can cause bruising, discolouration, crushing or tearing off of petals. Delicate flower stalks are vulnerable to physical damage.

9.1.4 Postharvest Loss Reduction Biotechnology

Harvesting, Packing and Storage

Gerberas are harvested using a sideward push near the base of the flower stem by holding the clump with the foot. The exact stage of harvesting varies with the grower, as well as with the distance from the market. The flowers are generally cut when the outer ray florets have completely elongated. The cultivars with flowers that close at night are harvested still later. It is beneficial to prick the stem 10 cm below the flower head to allow entry of water and also to remove the air from the cavity. The number of such cavities in the stem depends mainly on the cultivar, age of flower, light intensity at the time of lower development and the position of the cut stem.

In several western countries the flower heads are packed in insulated boxes to protect them from cold or freezing temperatures. Plastic-coated metal grids measuring 50×70 cm with a mesh size of 2×2 cm are used to pack *Gerbera* flower heads in Holland [12]. The individual flower head is supported by the grid, which is suspended above the plastic tray measuring 48×78×30 cm. The height of the grid can be adjusted according to the lenght of the flower stalks. The flower stalks can be shipped either dry or immersed in water or preservative solution in the tray. In Poland, returnable or non-returnable cardboard cups or mini sleeves (plastic) are used in place of metal grids to protect the cut blooms.

Gerberas are not suitable for long-term storage, as the flowers lose 40% of their vase life even after 7 days storage. For short-term storage, the optimum temperature is 1.65 °C [13].

Use of Floral Preservatives

The use of floral preservatives can double or even triple the postharvest lasting quality of cut gerberas. A solution containing sucrose and/or 8-hydroxy quinoline citrate, although it extends the vase life of flower heads, causes excessive elongation of flower stems. Stem elongation can be avoided by 10-min pulsing in 1000 ppm silver nitrate or 660 ppm sodium hypochlorite shortly after harvest. When silver nitrate or sodium hypochlorite is used, the stems should not be recut before placing in vase solution [9, 13–15]. Ag±EDTA complex has also been shown to be very effective in prolonging the life of cut gerberas. High fluorides in vase solution (0.2 ppm) cause brown spotting on petal tissues [13].

One of the main factors that influences the postharvest quality of cut gerberas is the cultivar being produced. According to Systema [14], cultivars like Pimpernel, Yellow Moor, Apple Blossom, Agnes, Terra Nigra, Symphonie, Carmen, Uranus, Terra Mix, Petrusa and Schlosser have the best keeping quality.

9.2 Anthurium

9.2.1 Introduction

Anthurium, flamingo plant" or "painter's palette" is a flower of increasing commercial importancce. In a floral arrangement, *Anthurium* flowers distinctly contribute to the elegance and attractiveness which are the prerequisites for a quality design. This flower has very good transportability [16, 28]. Of the many species, *A. scherzerianum* and *A. andreanum* are the most popular among florists for cut flower production. The showy portion of the flower head is one large heart-shaped spathe. The place wehre the rounded ends of the spathe meet has a very long style, sometimes straight, sometimes curved, and very evocative of the strange beak of the flamingo, hence the name flamingo plant. The spathe may be either flat or slightly undulated, ending with a pointed tip. The hermaphrodite flowers, which are small and insignificant, are densely packed on a cylindrical spadix with a large bract-like sphathe at the base.

Through continuous improvements by seedling selection and breeding programmes, today we have hundreds of seedling anthuriums, but only a few have acceptable qualities. An ideal plant should be a prolific producer of flowers. Apart from flower production, a desirable plant should grow vigorously and have shorter internodes.

A heart-shaped symmetrical spathe with overlapping or fused lobes is preferred by the public. The spadix should be shorter than the length of the spathe and recline artistically and gently in order to facilitate ease in packing. The size, shape, colour, and texture of the spathe are also important considerations in marketing of *Anthurium* cut flowers. The spathe should be free from sunburn, disease or other kinds of injuries. The flower stems should be straight, sturdy and should have shorter internodes.

Table 9.1. Standards for anthuriums[a] [30]

Grade	Average length and width of spadix (cm)
Extra large	Over 15.0
Large	12.5 – 15.0
Medium	10.0 – 12.5
Small	7.5 – 10.0
Miniature	Under 7.5

[a] The grades are recommended by the Hawaii Department of Agriculture

The double, blush, and tulip types of anthuriums are gaining in popularity. The Hawaii Department of Agriculture has suggested standard grades for *Anthurium* cut flowers [30] (Table 9.1.).

The present-day florist *Anthurium* are mostly hybrids of several species, as they seem to cross readily. Cultivars like Ozaki (and series), Chameleon, De Weese, Annenue, Red Elf, Avogerry, Avojok, Avo Claudia, Egmuntha, Haga White, Fulgo, Avotineke, Avonelly, Avocintha, Avoanneke, Sunburst, Avojose, Kaumana, Kozohara, Kansako, No. 1, Nakazawa, Uniwai, Marian, Seefurth, Nitta, Tateishi, Trindidad, Calypso, Horning orange, Horning, Rubin, Manoa Mist, Abe, Vogel Series, are very commonly grown both for cut flowers and in pots.

Both *A. andreanum* and *A. scherzerianum* undergo two distinct developmental stages, a juvenile phase (with only vegetative buds in the axil) and a generative phase. The flower buds produced in the axils of leaves during the generative phase remain dormant after initiation and develop into flowers depending upon the dormancy breaking. Unfavourable environmental conditions during this phase can kill the flower bud before normal development, or remain vegetative, depending on the stage at which the stress situation is created. The sequence – leaf, flower, and new leaf is maintained throughout the entire life of the plant. The interval between leaf production can be modified by manupulating the environmental conditions. The optimum temperature for vegetative growth is 18.3 °C. Lower growing temperatures cause spotting on leaves. Higher temperatures (21.1 to 23.9 °C) are needed to force the plant to initiate flower buds [30].

Some of the cultivars like Horning Orange, and Horning Rubin are neutral in their response to temperature (thermozero) [25]. However, the total number of flowers produced per plant is negatively correlated with rising air temperatures in the growin environment, especially in certian thermoneutral cultivars like Horning Rubin. Cultivars belonging to the Vogel group produce maximum yields of best quality flowers at 19 °C, while they do not survive at 10 °C and show leaf necrosis 13 °C [25].

Nature of Losses. Postharvest losses in anthuriums mainly result from mechanical injuries to spathe and/or spadix, mineral deficiencies or injuries caused by insect pests and diseases. The flowers start showing symptoms of wilting even at moderately mild water stress [22].

Physiological Disorders. Colour breakdown or discolouration of spathe, misshaped spathe and/or spadix are the most common physiological disorders in *Anthurium* flowers. Colour breakdwon in *Anthurium* has been associated with calcium deficiency. This disorder is very serious, especially when the pH of the substrate is below 4 [19–21]. Wilting of spathe caused by physiological stem plugging or high water stress can lead to huge losses in transit.

Pests and Diseases. Decline is another most common disorder in this crop. The symptoms of the disease include reduced size of plants, poor growth and lower flower yield. Nematode infection was found to be commonly associated with the symptoms [30].

9.2.2 Loss Reduction Biotechnology

Harvesting, Packing and Storage

Anthuriums should be harvested either manually or using harvesters when three-fourths of the flower (spathe) is developed [29]. Flowers harvested at a too early stage will have shorter vase life than those harvested at later stages. In some western countries, the development of true flowers (on the spadix) is used as a criterion for judging the harvest maturity. *Anthurium* growers in Hawaii harvest their crop when one-third of the true flowers on the spadix are fully developed [18, 30].

In one of the earliest methods of shipping cut anthuriums, the basal end of each flower stem was placed in a rubber balloon filled with water and the balloon was securely tied to the stem to supply water continuously. In another method, a cotton pad saturated with water and covered with wax paper was placed at the botton end of the flower stem and tied to the stem. In both methods, the flowers were then placed in a corrugated carton and the flower stems were secured properly to prevent their displacement during transportation. In the recent past, other methods, such as packing the flowers in plastic bags before placing them in cartons or dipping the spadix in melted parafin to reduce moisture loss, are also very popular in Hawaii and other North American states [17, 23]. The growers in The Netherlands follow the recommendation suggested by Segers [26]. With this method, the flower stems placed in flasks containing water are packed in moist boxes and soft protective material is put in between spathe and spadix. Foam plastic supports are provided in the box and the flowers are secured carefully with tape. The most commonly used box sizes for packing *Anthurium* flowers are $21.6\times50.8\times91.4$, and $27.9\times43.2\times10.16$ cm. These boxes can accomodate at least ten dozen flowers. Polyethylene lining provided in the box helps to maintain high humidity.

Low Temperature Storage

Anthurium can be stored for up to 3 to 5 weeks at 13.0 °C [13]. Exposure of flowers to temperatures below 13 °C causes chilling injury, as indicated by darkening or blueing of the spathe. Dark red coloured flowers are most susceptible to chilling injury. A postharvest dip of 10 ppm of N^6BA can decrease respiration, increase vase life and reduce chilling injury [27].

Controlled Atmospheric Storage

In the absence of refrigeration, controlled atmospheres created by 2–10% O_2 can be advantageously used to hold the flowers at ambient temperatures [18] (24–25 °C).

The cultivars differ in their postharvest lasting qualities. Certain cultivars have been shown to have up to 60 days of postharvest life, while a few others have only 2 days [13].

Use of Floral Preservatives

The use of preservative solution is beneficial at all stages in the marketing channel. A preshipping dip of flower stems for 10–60 min in 4 mM silver nitrate solution (within 12 h after harvest) has been found to extend the vase life following simulated shipping by 40–60% [24]. Several commercially available floral preservatives (Floralife, Everbloom, Roselife) and chemicals (sodium benzoate, benzoic acid, glucose, sodium hypochlorite, hydrochloric acid and carbonated beverage) can also be used to prolong the shelf life of *Anthurium* flowers [17].

Control of Pests and Diseases

Prevention of diseases and pests through regular spraying schedule is essential to maintain proper hygiene and to produce best-quality flowers.

9.3 Gypsophila

9.3.1 Introduction

Gypsophila (*G. paniculata*, family Caryophyllaceae) is named for its ability to grow in calcareous soils (gypsum-loving plants). In the florist trade, they are commonly known as baby's breath. The inflorescence in this plant is a multibranched panicle. The flowers at the top open first and flower opening proceeds basipetally. *Gypsophila* flowers can be dried and stored for about a year, while the vase life of fresh baby's breath inflorescence is only about 5 to 7 days. White *Gypsophila* is most commonly sold on the market. However, other colours, like pink and red, are also available in the trade.

Flowers dry very rapidly after harvest. Low relative humidity in the storage environment drastically reduces subsequent lasting qualities. Another major problem in this crop is that the flowers do not open simultaneously. This makes it difficult to determine the ideal stage for harvesting. The postharvest keeping quality is influenced by the stage of harvesting.

Traditionally, the sprays are harvested separately as soon as the flowers at the top open. The stems are cut 25 to 37.5 cm long and are immediately placed in water or preservative solution and kept in the refrigerator to avoid drying of flowers. With the introduction of bud harvesting or once-over harvesting, the inflorescence can be cut in the bud stage and is opened in preservative solutions [32]. The most common bud-opening solution for *Gypsophila* contains 2% sugar+200 ppm 8-HQC. A solution containing 5–10% sucrose and 25 ppm silver nitrate can also be used with similar effects [13]. Flower opening was found to be improved by bud-harvest technique.

Cut gypsophilas in water and under low moisture stress (high relative humidity) conditions can be stored at 4.4 °C for about 2 days. A solution containing 25 ppm silver nitrate (not registered) and 5 to 10% sucrose has been recommended by Staby et al. [13] for low temperature storage. As far as possible, dry storage must be avoided.

Baby's breath flowers are shipped by air in standard cardboard boxes. However, it is very important to make sure that the insulation in the box is adequate. Preshipping conditioning of flowers in 5 to 10% sucrose and 25 ppm silver nitrate (not registered) solution for over night is beneficial in improving the lasting qualities of cut *Gypsophila.* These flowers, on arrival in the retail shop, should be unpacked immediately, placed in water, and held under refrigerated conditions to avoid drying of individual flowers. Use of floral preservatives containing surcrose (5 to 10%) and silver nitrate (25 ppm) is beneficial in extending the life of cut *Gypsophila* flowers. Everbloom, modified Cornell solution or 100 ppm sodium benzoate (not registered) along with 8-HQC and sucrose can also be used for this purpose [13]. Under ideal conditions, cut *Gypsophila* can last for 5 to 7 days.

9.4 Iris

9.4.1 Introduction

The present-day iris (*Iris* spp., family Iridaceae) cultivars have been derived from intercrossing among *I. xiphium, I. tangitania, I. lusitanica,* and *I. filifolia*. The Dutch iris (*I. hollandica*), due to longer flower stalks, is primarily used for cut flower production. It represents the cross among *I. xiphium, I. tangitania,* and *I. lusitanica* [33]. Three cultivars, Wedgewood, Ideal and Prof. Blaauw (Blue Ribbon) are the most common Dutch iris produced in the United States, The Netherlands and France. The former two cultivars are suitable for an early forcing season while the latter is very good for mid- or late-season forcing. Other types, like German, Japanese, Apogan, dwarf and Oncocylus, are also important from the horticulturist's point of view. Some of the dwarf types are used for pot culture.

Several factors affect growth, flowering and flower quality in bulbous crops. Important ones include bulb size, environment (light, temperature, humidity and ethylene pollution), leaf formation, the stage of flower development at the time of harvest and low temperature treatment. For flowering Dutch irises, bulb size of at least 8 to 12 cm is recommended to produce quality flowering [33]. The number of leaves present in the bulb influences the flowering in this crop. A bulb having less than three leaves before flower initiation will not produce flowers. In the case of Dutch irises, insufficient light (less than 100 lx), will cause flower bud abortion [34]. Flower bud initiation takes places during low temperature treatment [33].

Losses in quality and quantity of cut irises can be related to several factors, such as suboptimal conditions during bulb storage (programming) and forcing which cause blindness or blasting, use of lower-grade bulbs for forcing or disease and pest attack. Blindness is realted to the use of smaller-sized bulbs for forcing, early forcing of bulbs, or improper programming. Flower bud blasting or flower bud abortion is also very common in Dutch irises and is caused by low light intensities, high temperatures, improper fertilization, water stress or higher level of ethylene during storage and forcing.

9.4.2 Postharvest Biotechnology

The optimum stage of harvesting depends on the cultivar and the distance from the market. The Wedgewood iris should be cut when the flower bud is just enough to show the colour. Flowers harvested prior to reflex bending of outer tepals have been shown to last longer than those harvested in later stages [13]. In contrast, Blue Ribbon iris harvested when the tepals open (reflex bending) perform better in the postharvest environment. For long-distance transport, partially opened flowers are more suitable, as they can be packed much better than those that are fully open. Flowers can be allowed to open on the plant before harvest for local sales.

The removal of basal plate encourages better water uptake by the cut flower in this crop. Conditioning of flowers with warm water or preservative solution containing 200 ppm 8-HQC, 50 ppm silver nitrate (not registered) and 5% sucrose is beneficial in improving their postharvest performance. Preconditioning with floral preservative has been shown to improve the colour in flowers [13].

The flowers of iris can be stored for a longer period under low temperatures (−0.56 to 0 °C). For extended storage, the flowers should be kept in water and should always be in upright position to avoid tip bending.

The flowers, on arrival in the retail shop should be conditioned in warm (32–42 °C) water for 2 to 4 h prior to use [13]. Recutting the stems before placing them in water is beneficial in improving water conductance. Use of floral preservative solution in bulbous crops in general is still debatable, with some research workers advocating it [13] while others do not [33].

9.5 Narcissus

9.5.1 Introduction

Narcissus (*Narcissus* spp., family Amaryllidaceae) is a well-known spring flowering bulb. Narcissus (daffodils) are plants with truncate bulbs. The leaves are produced with white or yellow blooms, which appear solitarily on a longe peduncle the spathe is one-leaved, membranous, and the perianth tube varies in shape (ascceding, spreading or reflexed). The coronar or crown is long, tubular or cup-shaped and is attached to the perianth tube.

The important daffodils in the trade are *N. jonquilla*, (jonquils), *N. tazetta* (Paper white), *N. poetaz*, *N. poeticus* and *N. bulbocodium*. These species very in their requirements for sprouting and growth, and take varying lengths, of time for anthesis.

The bulbs are sold based on number of noses, the double-nose No. 1 size will produce about two flowers per bulbs, while the double-nose No. 2 gives 1.5 flowers per bulb [35]. Unlike tulips, the daffodil bulbs have all the flower parts formed, but they are poorly differentiated. Therefore, the programming in this crop should allow complete flower formation prior to the normal low temperature storage of bulbs. Warm-cool-warm temperature sequence is followed to ensure proper programming in daffodil bulbs [33]. Forcing conditions greatly influence flower size and keeping qualities. For cut daffodils, the plants are forced at 13 – 15 °C [33]. Lowering the temperature to 12.7 °C just prior to full anthesis improves the keeping quality of flowers [35].

Postharvest quality or the performance of cut daffodils is influenced by several factors like bulb storage, forcing conditions, harvesting stage and handling of flowers after harvest. Prolonged storage of bulbs at warm temperatures (beyond 4 days at 31.9 °C) drastically reduces stem length and flower size. At the same time, an insufficient or too short precooling period also affects lasting qualities in cut daffodils. The cultivar differences for bulb forcing should also be recognized.

Greenhouse forcing conditions are equally important for the production of top-grade flowers. Excessively high temperatures and light intensities, particularly at later stages of plant growth, are most detrimental. Under high light intensity conditions, the stretching of stems is a major problem. This results in weak stems and affects the overall quality of flowers. Similarly, high temperatures, especially during flower bud development, result in failure of the flower to grow beyond the "goose neck" stage. This disorder is comparable to flower abortion, in that it is flower abortion that occurs at later stages and is commonly referred to as "bull nosing". Bulbs infected with Fusarium (basal rot) produces very poor growth and unmarketable flowers.

Loss reduction methods must always begin with handling of bulbs and be continued through programming, greenhouse forcing of the crop, harvesting and handling of finished products at different stages in the marketing channel.

The stage of harvesting is determined based on the development of flowers. Daffodils are traditionally cut in their "goose-neck" stage (Fig. 9.1). Some

Fig. 9.1. Goose-neck stage of flower in daffodil [33]

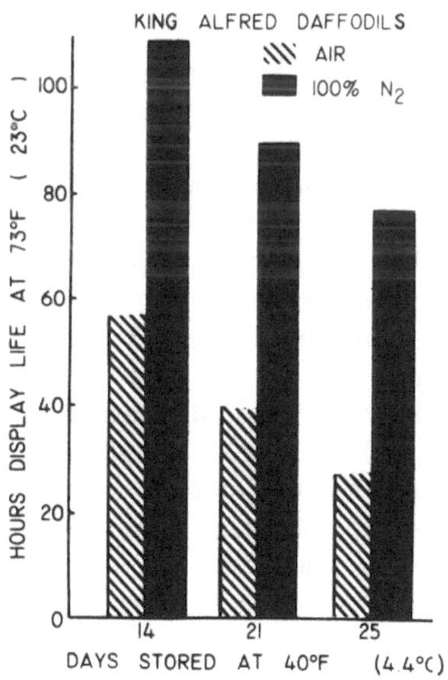

Fig. 9.2. Modified atmospheric storage of daffodils [36]

growers, with better talent and facilities, harvest their daffodil crop as soon as the colour shows up on the flower bud. In general, the flowers harvested later than the "goose neck" stage do not perform well under postharvest conditions.

The flowers of daffodils can be stored for 3 weeks at 0 to 0.6 °C without any loss in quality [13, 33]. Flowers kept in water or preservative solutions (preferred) were found to be better than those held under dry conditions. For long-term low temperature storage, flowers should be held in upright position to avoid bending of stems.

Cut daffodils can also be stored in 100% nitrogen atmosphere without any adverse effects on flower quality. At 0 °C the flowers held in 100% nitrogen had better display life (103 – 123 h) than those held in normal air (85 – 98 h) [36] (Fig. 9.2).

Because of the phytotoxic effects of 8-HQC on flowers use of flower preservative containing more than 150 ppm of 8-HQC is not recommended in this crop. However, use of floral preservative can extend the keeping quality of cut daffodils by up to 3 days [13]. Conditioning of flowers for 2 to 4 h with warm water (32 – 42 °C) at room temperature immediately on arrival in the retail shop is also beneficial in improving the postharvest performance.

9.6 Snapdragon

9.6.1 Introduction

Snapdragon (*Antirrhinum majus* L.) is a relatively recent introduction in the greenhouse. In the past, they were grown either as a perennial or annual for summer flowering in flower beds. The commerical production for cut flower purposes began with the introduction of a winter flowering cultivar Cheviot Maid. This followed the introduction of several F_1 hybrids having desirable cut flower qualities.

The flowers of snapdragons are sold throughout the year for use in floral arrangements and for bouquets. The yellow and pink snapdragons are highly preferred by the public as compared to the other colours. Most snapdragons are grown as single crops, but in some cases two or three crops can be taken before they are discarded. With the introduction of some of the F_1 hybrids, it is now possible to grow them all the year round.

The postharvest quality attributes in snapdragons include stem length, total number of florets, number of open florets, weight to spike, freedom from abnormal growth, crooked stems, pests and diseases, physiological disorders, vase life, and retention of petal colour [37, 39]. The minimum specifications recommended by the Society of American Florists are given in Table 9.2. Sometimes the term "unclassified" grade is used in the snapdragon trade to designate the lots which are not graded according to established standards [37].

Laurie and Poesch [40] were the first to show that flowering in snapdragon is promoted by long photoperiods. This was later confirmed by other investigators in several cultivars [41 – 46]. In contrast, Maginnes and Langhans [43] suggest that snapdragon's response to photoperiod was a quantitative effect. The stage of development of plants at the time of photoperiodic treatment was found to influence their response to day length. The plants which were

Table 9.2. Specifications of SAF standards for snapdragons [37]

Grade	Label colour	Weight per spike (g)		Minimum open flowers per stem	Minimum stem length (cm)	Stems per bunch (Number)
		Minimum	Maximum			
Special	Blue	71	113	15	91	12
Fancy	Red	43	70	12	76	12
Extra	Green	29	42	9	61	12
First	Yellow	14	28	6	46	12
Utility[a]	White					

[a] The utility grade is not a grade by itself. The cut flowers that are not classifiable into the regular grades due to abnormal growth, foliage injuries, curved stems, stem breakage, or any other abnormality are included in this grade

5–20 cm high showed maximum response to long days [42]. Other scientists suggested the 10- to 12-leaf stage (or 5 to 7 weeks from sowing) to be the most sensitive [43].

Like roses, chrysanthemums and carnations, snapdragons need full sunlight for optimum growth. Reduction in light due to overcrowding of plants has been shown to affect the cut flower quality to a great extent.

Senescence of cut flowers of snapdragon is brought about by self-generating ethylene. Snapdragons have been known to produce a large quantity of ethylene to gas themselves [52]. Floret abscission occurs within 24 h after exposure to ethylene (Fig. 9.3). The response to ethylene varies with the cultivar, in-

Fig. 9.3. Response of snapdragon cultivars to self-generated ethylene during dry storage at −0.56 °C in a closed container. Picture take 24 h after removal from storage. *Left* cv. Christmas Cheer; *right* Dorcas Jane [81]

dicating varietal differences in the amount of autocatalytic ethylene produced in the flower tissue. Diseased leaves also serve as a site of ethylene biosynthesis for autocatalytic ethylene production [53]. In general, ethylene concentration above 0.5 ppm in the atmosphere causes floret abscission within 24 h [88].

9.6.2 Postharvest Losses in Snapdragon

Postharvest losses in cut snapdragons are attributed primarily to self-generating ethylene gas which causes flower and leaf abscission. The leaves infected with pathogens such as *Puccinia antirrhini* can act as an ethylene source and stimulate ethylene synthesis by other flower tissues.

Drastic reduction in cut flower quality in snapdragons may result from stem tip breakage or bending, abnormal stem development, crooked stem, and foliage injury. These factors reduce the mean flower grades. Postharvest losses in both quality and quantity of snapdragons are related to cultural [41–51], mechanical [38], physiological [52–64] factors and attack of pests and diseases [53]. The photoperiod, together with growing temperatures, have tremendous influence on flowering and flower quality in shapdragons [41–51].

Boron nutrition is very important in snapdragon production [54, 63]. Boron deficiency causes abortion of the terminal growing point or flower stem tip. This results in growth of side shoots and drastic reduction in the number of florets produced on the main stem. The problem of boron deficiency becomes more serious in calcium-rich soils.

Ethylene gas causes premature floret drop (shattering). Exposure of plants to 0.5 ppm or more ethylene was found to be enough to cause floret shattering or "shelling" within 24 h [64]. This response in snapdragons has been shown to be controlled by a single dominant gene [64].

Stem tip breakage is a major problem in snapdragon handling. It is caused by poor lignification in stems. The point at which the stem breaks has been shown to be a varietal character. For example, red and rose cultivars break very close to the terminal point, the bronze and yellow cultivars farther away from the apex. Adams and Urdahl [65] correlated this phenomenon in bronze and yellow cultivars with the greater demand for phenyl propanoid precursors needed for flower pigment formation in these cultivars. Lignification was shown to be a function of the amount of phenyl propanoid precursor [65].

Geotropic bending of the flower spike is another problem that is quite often noticed in snapdragons during transportation. The bending occurs with a few minutes of placing the cut flowers horizontally and is almost irreversible. The cut flowers with bent stems are unsuitable for grading and often of limited use in floral arrangements or in the vase.

The severity of the various diseases largely depends upon the location and the growing conditions. Some of the more common diseases include *Botrytis* blight, powdery mildew and downy mildew. Other pathogens like *Alternaria* and *Helminthosporium* are also noticed on flowers of snapdragons. In some parts, snapdragon leafspot is a very serious problem. The infected tissues have been shown to produce a large quantity of ethylene gas.

9.6.3 Loss Reduction Biotechnology

Snapdragons are very responsive to postharvest treatments. It has been well established that flowers produced under optimum growing conditions perform better is postharvest environments. This demands proper attention to forcing the plants under right light and temperature conditions, adequate nutrition, timely control of pests and diseases, proper stage of harvesting, grading, packaging, transportation and the use of chemical growth regulators and other chemicals to prevent shelling or shattering of florets.

Harvesting, Grading and Transportation

The spikes are harvested when the lower one-third of the florets is fully open. More recently, some of the growers having facilities for opening of florets in preservative solutions harvest their crop when the lower one or two florets are open. It is important to make sure that the remaining florets on the spike can successfully develop under postharvest conditions. The spikes should be kept in vertical position at all times to avoid geotropic bending [66–68].

Another major problem in handling cut snapdragon spikes is stem-tip breaking which occurs at the end of xylem lignification [37, 65]. Since this phenomenon is genetically controlled in snapdragons, evolving varieties with resistance to stem breaking will help to solve this problem.

Shelling or shattering of florets in response to ethylene in the atmosphere is yet another serious problem that makes handling of cut spikes more difficult. This necessitates proper precautions to remove ethylene in the storage or holding environment. In recent years, plant breeders have been trying to evolve varieties resistant to ethylene-induced floret shattering. Since the degree of sensitivity to ethylene varies with the cultivar, proper selection of cultivars for forcing may help to reduce losses due to ethylene-induced shelling of florets.

The flowers are normally graded into special, fancy, extra, first or utility groups based on criteria given in Table 9.2. The grades recommended by the Society of American Florists are only voluntary and the growers are not bound by these standards. Several growers use their own grading, based on experience. However, the top grade flowers must have straight, strong stems, free from any abnormality, clean, uniform and healthy foliage, and flowers representative of the cultivar. The florets must be fresh, clean and free from pests, diseases, physical and physiological injuries. Traditionally, foliage from the lower one-third portion on the spike is removed to avoid the leaves coming into contact with the vase solution and rotting.

Snapdragons are not generally transported over a long distance. If they are to be shipped, then they are kept in upright position in hampers or containers to avoid tip bending. In recent years, some of the chemical growth regulators are being used to check geotropic curvature [38, 68, 70].

Low Temperature Storage of Cut Snapdragons

The recommended storage temperature for snapdragons is between $-0.55\,°$ and $1.65\,°C$ [64]. Based on the method of handling there are mainly two kinds of storage. In the first type, the cut spikes are either sleeved in paper bags or stored as such in dry packs for about 3 to 4 days without any appreciable loss in subsequent lasting qualities. In the other method, which is suitable for extended storage, the flowers are stored in water or preservative solution with or without light. When the flower stems are immersed in only water, the storage under dark was slightly better compared to storage under light (200 ft. c) [68]. The stage of harvesting is important for long-term storage. Cutting the spike when the lowermost floret began to open was found to be better than the later stages, because the florets opened during storage have a faded colour.

Hypobaric Storage of Cut Spikes

The use of hypobaric conditions for storage of cut flowers was advocated for the fist time by Burg [69]. He found that cut spikes of snapdragons could be stored in low pressure (40 mm Hg) for 42 to 56 days without any reduction in subsequent lasting qualities. In contrast, they could be stored only up to 14 days under conditions of normal pressure. This response of snapdragons to hypobaric conditions is attributed primarily to the constant removal of ethylene from the storage atmosphere and the control of ethylene biosynthesis in flower tissues. Although the benefits of low pressure storage are well established, it is not a common practice to store flowers in hypobaric storage.

Use of Chemical Growth Regulators and Floral Preservatives

Snapdragon flowers react very favourably to postharvest chemical treatments [70–74]. In certain cases, the keeping quality of flowers was extended by two to three times. An ideal floral preservative solution should contain 300 ppm 8-HQC and 1.5% sucrose. Other scientists have recommended the use of 10 to 50 ppm of B-Nine (N-dimethyl amino succinamic acid) [70] or 50 to 100 ppm of cycocel in the preservative solution (300 ppm 8-HQC+1.5% sucrose) [68], 0.5% sucrose+300 ppm 8-HQC [73], analogues of rhizobitoxin and sodium benzoate [74]. Anthocyanin development was found to be greatly checked under dark conditions. Light intensity of 2150 lx was found to be adequate for proper colour development.

Control of Pests and Diseases

Proper precautions to avoid the occurrence of diseases and pests is absolutely essential to produce a quality crop. Maintaining adequate sanitation in the production area helps in removing potential inoculum sources and preventing occurrence of diseases.

Some chemicals, such as Malathion, nicotine, tetraethyl pyrophosphate (TEPP), sulphur and chlorobenzide have been shown to be injurious to snapdragons. These chemicals should be avoided for preventing spray injuries.

9.7 Statice

Two types of statice, notch-leaf statice (*Limonium sinnatum*), and German statice (*L. latifolium*) are commonly found in the florist's trade. These plants flower almost throughout the year, except for a short period in November. The flowers are produced in panicles with loose clusters. The most common flower colours in notch-leaf statice are shades of orange, yellow, pink, red, purple, lavender, blue and white. In the case of German statice, the mauve colour is the most common. Individual flowers are small and are borne on 30–45-cm-long-stalks. In the United States, statice is grown outdoors in Florida and California and the flowers are harvested in early summer. Cut statice flowers are used either fresh or dry. In dried condition, these flowers can be stored for over 1 year.

One of the major postharvest problems in cut statice is the growth of microorganisms in vase water, causing rotting of stems. The stems, especially in German statice, are very hard and need treatment to loosen them.

The flowers are harvested when they show colour [31]. They are then bunched in groups of ten stems. They should be placed in water or preservative solution after harvest and held at 4.4 °C to prevent premature wilting. Statice flowers can be stored at 1.7 to 2.0 °C for 2–3 weeks [13]. The preservative solution or water in the holding container should be changed at weekly intervals to prevent bacterial growth. For an extended period of storage, the flowers are dried in bunches by hanging them in a dry, well-aerated room for several days. These dried flowers should be stored in dry conditions, as contact of stem with moisture can render them useless.

On arrival in the retail shop, the fresh flowers should be unpacked immediately, stems are recut (lower 2 to 4 cm), and conditioned in warm water (38–44 °C). The damaged leaves and flowers must be removed before conditioning. In the case of German statice, the stems are recut and immersed in hot water for 1 min to loosen tough stems, and then handled in the normal way.

9.8 Freesia

Freesia (*Freesia refracta*, family Iridaceae) a native of South Africa, is the most popular cut flower in Europe and ranks next only to carnations. It is a perennial grown from corms. While Ballerina is the most common cultivar in The Netherlands and other parts of Europe, Fantasy Aurora and Blue Heaven are also grown to a certain extent [75]. The flowers of *Freesia*, which are borne on a spike, are produced at right angles to the main stem. There are up to eight florets in a spike which are white, yellow, blue, red or pink in colour.

Freesias are propagated through seeds and corms, with the former method having gained in popularity in recent years. When started from corms, they take approximately 5 months to bloom. As in other bulbous crops, the forcing technique influences the growth of plants, the number of florets per spike and also the quality of individual florets. The plants should be forced at a minimum temperature of 16 °C until four leaves are visible, after which the temperature is lowered to 13 °C.

The postharvest losses in freesias result from mechanical damage, physiological disorders and pests and diseases. Improper forcing can drastically reduce the quality and quantity of flowers produced. Diseases such as *Botrytis* blight, *Fusarium* yellow, *Stromatinia* rot, white break and viral diseases affect the production and quality of flowers.

Postharvest losses, in both quality and quantity, can be reduced by using healthy planting material (preferably seed to avoid viruses), adapting an appropriate forcing technique, harvesting at proper stages, and careful postharvest handling of flowers. The flowers are harvested when the lowermost floret shows colour. Best results can be obtained by harvesting the flowers in the evening and allowing them to stand in water overnight [77]. Then the flowers are bunched either manually or by using electronic machines [78]. These bunches are packed in plastic-coated boxes. The flowers can be stored dry for 5 to 7 days at 0° to 1.0 °C [76].

9.9 Delphinium

9.9.1 Introduction

Larkspur (*Delphinium* spp.) is a herbaceous hardy perennial in which flowers are arranged in long (0.5 to 1.5 m) racemes. The individual floret is 0.5 to 1.0 cm in diameter and comes in various shades of blue, pink, white and purple. Delphiniums are commercially propagated through seeds, terminal cuttings or by division of old clumps. The greenhouse forcing schedule involes growing the plants at 4° to 7 °C for 6 weeks, after which the temperature is raised to 10 °C for flower production. The crop is usually started in August when grown from seed and moved into the greenhouse in November. The low temperature (4 °C) should be started from February 1 for producing flowers between May and July.

Floret shedding accounts for most major postharvest losses in larkspur. The presence of ethylene and darkness in the holding environment increase flower shedding. Disease and pest infestation can also drastically reduce the quality and quantity of the crop.

In general, the spikes are cut when one-half to two-thirds of the florets have opened. They are handled similarly to snapdragons or mathiola. At 4 °C the spikes can be stored for 1 to 2 days [13].

9.10 Stocks

9.10.1 Introduction

Stock (*Mathiola incana*) bears terminal clusters of white, pink, red, purple or yellow flowers, which are either single or double. The stem length varies from 30 to 48 cm. The florist's stock is a non-branching and columnar raceme.

Stock is propagated through seeds, which take about 2 weeks for germination. The greenhouse forcing schedule involves growing the seedling at 16 °C until at least eight pairs of fully developed leaves are produced, after which they are maintained at 10 °C for 3 weeks for flower bud initiation. Best-quality flowers are produced by raising the growing temperature to 16 °C (day) with a night temperature of 2 to 4 °C after complete flower initiation [78]. In general, 9 weeks are needed for complete flowering.

The postharvest losses in stock result from mechanical damage, floret abscission, tip bending, and diseases and pests. *Botrytis* rot and stem rot are the most common diseaes, causing huge postharvest losses.

Like snapdragons, spikes are cut when the lowermost one-half to two-thirds of the florets have opened. These flowers need to be maintained in water or preservative solution. The flowers can be held for 3 to 4 days at 4 °C [13].

9.11 Zantedechia

9.11.1 Introduction

Calla lily or lily of the valley is a tropical perennial with solitary showy flowers. The individual flower has a showy spathe and a spadix. *Zantedechia aethiopica* has a fragrant white spathe 15 to 20 cm wide which tapers to a point, while *Z. elliottiana* has a trumpet-shaped yellow spathe and *Z. rehmanii* has a pink-coloured spathe. All of these species are propagated by rhizome. Forcing the plants of *Z. aethiopica* at 13 °C and that of *Z. ellittiana* and *Z. rehmanii* at 16 °C produces best-quality flowers.

The flowers are harvested just before the spathe beings to bend downward. The flowers are handled similarly to anthuriums. At 4 °C, these flowers can be maintained for 7 to 10 days [13].

9.12 Paeonia

Paeonia (*P. lactiflora* and *P. albiflora*) is a hardy perennial with large soiltary flowers. The flowers, which are up to 25 cm wide, have several petal that are pure white, yellow, red or pink. Peonies are grown outdoors during cold weather. The crop is grown by division in early fall. The flowers are cut when the buds start showing colour. The peonies harvested in the bud stage can be stored dry for 4 to 6 weeks at 0–2 °C [13].

9.13 Anemone

Anemone, or lily of the field (*A. coronaria*) is a perennial having solitary flowers with petal-like sepals on medium to long (30 to 45 cm) stalks. Individual flowers, which are single, double or semi-double, come in shades of blue, red, pink, lavender and white. The seed-grown anemones are preferred to those grown from tubers, on account of diseases.

The crop is started from seeds in early spring. High growing temperatures usually reduce the flower size and stem length. A growing temperatures of 4 to 7 °C is found satisfactory in producing quality flowers.

Flowers are cut in the bud stage and can be stored for 1 to 2 days at 4 to 5 °C.

9.14 Aster

China aster (*Callisteprus chinensis*) a flowering annual, is valued for its wide range of colour, form and size of flowers. The solitary flowers, which are 5 to 10 cm in diameter, come in shades of blue, purple, pink, red, and white. Asters are grown both outdoors and in greenhouses. The crop is usually started from seed during relatively mild weather (night temperatures around 10 °C). The plants take about 120 to 150 days to produce flowers. The flowers are handled similarly to spray mums.

9.15 Rananculus

Rananculus (*R. asiaticus*), a herbaceous perennial, is also known in the trade as Persian buttercup or turban. One to four white, yellow or pink flowers are present on each stem, which are about 20 to 30 cm tall. Rananculus is propagated through seeds or division of tubers. The flowers are handled similarly to anemone.

9.16 Sterlitzia

Birds of paradise (*Sterlizia reginae*) are valued for greenish purple boat-shaped bracts which contain the true flowers. The flowers are suitable for flower arrangement, and have orange or reddish orange outer petals and blueish inner petals. These petals, 7 to 10 cm long, are pointed at the tip. Sterlizias take nearly 3 years to flower. Forcing the crop under partial shade at 10 to 13 °C (night) with 20 to 22 °C during the day produces the best quality flowers.

The flowers are cut when the first floret opens. However, the postharvest life can be extended up to 1 month by harvesting them in the bud stage and pulsing for 2 days at 22 °C in a solution containing 8-HQC, sucrose and citric acid. The flowers can be held for 4 days at 10 °C. Long-term storage is not normally recommended for sterlizia [13].

9.17 Cut Greens

The common florist's cut greens include *Asparagus* spp. (Smilax and Plumosa fern(, *Buxus sempervirens* (boxwood), *Camellia japonica, Codiaeum variega-*

Table 9.3. Storage requirements for some florist's cut greens

Crop	Storage requirements
Asparagus spp.	Store in water at 2° to 4 °C, Preservatives reduce the keeping quality changing the vase water once a week is beneficial
Buxus sampervirens	Store in preservative at 5 °C
Dryopteris erythrosora	As above
Camellia japonica	As above
Cytisus scoparius	As above
Croton	Store in preservative at 2° to 4 °C
Cordyline terminalis	Store at 4 °C for 2 to 3 weeks. Use of water or preservative
Euclytus sp.	As above
Gaultheria shallon	Store at 0 °C, maintain high relative humidity
Ilex spp.	As above
Hedera spp.	Store in preservative at 2° to 4 °C
Pittosporum tobira	Store at 4° to 7 °C in preservative
Podocarpus macrophyllus	As above

tum, Dracaena terminalis, Cytisus scoparius, (Scotch broom), *Euclyptus pulverulenta, Gualtheria shallon* (lemon leaf), *Ilex aquifolium, Pittosporum tobira,* and *Podocarpus macrophyllus.* Under favourable postharvest conditions, these cut greens last for 1 to 3 weeks. Thes plants are multiplied either by cuttings or by division.

The condition of the foliage and the stem determines the quality of cut greens. Yellowing, rotting, drying and abscission of leaves cause considerable postharvest losses in cut greens. Disease and pest infestation also reduce the consumer acceptability for these products. Sheeding of spores, as in *Pteris cretica, Cyrtomium falcatum*, or shrivelling fronds, as in sword fern, can also ruin the quality of cut greens.

Best results could be obtained by maintaining cutgreens under light with their stems immersed in presentative solution or tap water [79, 80]. Cutting about 1 cm stem every day has been found to improve the postharvest lasting qualities.

Storage requirements for some florist's cut greens are given in Table 9.3.

References

1. Byrne TG (1973) Gerbera selection and postharvest handling. Flower Nurs Rep Nov 10
2. Leffring L (1973) Flower Production in Gerbera-I. Correlation between shoot, leaf and flower formation in seedling. Sci Hortic 1(3):221
3. Plate HP, Krober H (1975) Damage to young gerbera plants caused by *Pythium* sp. and *Rhizoctonia solani.* Gartenwelt 75(15):340
4. Leffring L (1973) The influence of leaves and shoots on flower production in *Gerbara.* Vakbl Bloemisterij 28(15):11
5. Meeteren van U (1978) Water relations and keeping quality of cut *Gerbera* flowers II. Water balance of ageing flowers. Sci Hortic (Amst) 9:189
6. Meeteren van U (1979a) Water relation and keeping quality of cut *Gerbera* flowers. III water content, permeability and dry weight of ageing petals. Sci Hortic (Amst) 10:261

7. Meeteren van U (1979b) Water relations and keeping quality of ageing cut *Gerbera*. IV Internal water relations of ageing petal tissues. Sci Hortic (Amst) 11:83
8. Meeteren U van, Gelder H van (1980a) Water relations and keeping quality of cut *Gerbera* flowers V. Role of endogenous cytokinins. Sci Hortic (Amst) 12:273
9. Meeteren U van (1980b) Water relations and keeping quality of cut *Gerbera* flowers. VI. Role of Pressure potential. Sci Hortic (Amst) 12:283
10. Meeteren U van (1980c) Role of pressure potential in keeping quality of cut *Gerbera* inflorescences. Acta Hortic 113:143
11. Saccalis JN (1975) Sucrose: patterns of uptake and some effects on cut flower senescence. Acta Hortic 41:45
12. Gradner U (1975) A palette for transporting gerberas developed at Veitschochheim. Dtsch Gartenbau 29(42):1943
13. Staby GL, Robertson JL, Kiplinger DC, Conover CA (1978) Chain of life, Ohio Flor Assoc Columbus, Ohio (USA)
14. Systema W (1981) The keeping quality of gerbera varieties is difficult to assess. Vakbl Bloemisterij 36(4):50
15. Veen H (1983) Silver thiosulfate an experimental tool in plant science, Sic Hortic 20(3):211
16. Akamine EK (1976) Postharvest handling of tropical ornamental crops. HortScience 11(2):125
17. Akamine EK, Goo T (1975) Vase life extension of anthurium flowers with commercial floral preservatives, chemical compounds and other materials. Flor Rev 155(4027):14
18. Akamine EK, Goo T (1981) Controlled atmosphere storage of anthurium flowers. HortScience 16(2):206
19. Higaki T, Rasmussen HP, Carpenter WJ (1980) Calcium deficiency of *Anthurium andreanum* Lind. Spathe. J Am Soc Hortic Sci 105(3):438
20. Higaki T, Rasmussen HP, Carpenter WJ (1980) Color breakdown in *Anthurium andeanum* Lind. spathe caused by calcium deficiency. J Am Soc Hortic Sci 105(3):441
21. Iwata RY, Tung CS, Kamimoto H (1979) Anthocyanin of *Anthurium andreanum*. J Am Soc Hortic Sci 104(4):464
22. Kamemoto H (1962) Some factors affecting the keeping quality of anthurium flowers. Hawaii Farm Sci 11:2–4
23. Nijlunsing W (1976) Transport trial with anthurium packaging. Vakbl Bloemisterij 31(42):23
24. Paull RE, Goo T (1982) Pulses treatment with silver nitrage extends the vase life of *anthurium*. J Am Soc Hortic Sci 107(5):842
25. Schenk M, Brundert W (1981) Temperature effects on *Anthurium andreanum* hybrids. Dtsch Gartenbau 36(49):2064
26. Segers CM (1976) Pack anthurium flowers more carefully. Vakbe Bloemisterij 31(45):45
27. Shirankawa T, Dedolph RR, Watson DP (1964) N^6-Benzyl adenine effects on chilling injury, respiration and keeping quality of *Anthurium andreanum*. Proc Am Soc Hortic Sci 85:642
28. Systema W, Barendse L (1975) Keeping quality of cut flowers VI. *Anthurium andreanuma* cut flower that lasts well. Vakbl Bloemisterij 30(51):1921
29. Walker PN, Smith MR (1978) Machine logic for harvesting of anthurium harvester. Transaction of ASAE 21(6):1040
30. Higaki T, Watson DP, Leonhardt KW (1979) Anthurium culture in Hawaii. Univ Hawaii Manos Circular 420
31. Auman CW (1980) Minor cut flower crops. In: Larson RA (ed) Introduction to Floriculture. Academic Press London 185-211
32. Farnham DS, Kofranek AM, Kabota J (1978) Bud opening in *Gypsophila paniculata* cv. Perfecta with Physan 20. J Am Soc Hortic Sci 103(3):382
33. De Hertough AA (1980) Bulbous crops. In: Larson RA (ed) Introduction to Floriculture. Academic Press, London 215
34. Fortanier EJ, Zevinbergen (1979) Analysis of the effects of temperature and light after planting on bud blasting in *Iris hollandica*. Neth J Agr Sci 21:145
35. Laurie A, Kiplinger DC, Nelson KS (1968) Commercial Flower Forcing. McGraw-Hill, New York
36. Hardenberg RE, Uota M, Parson CS (1969) Refrigeration and modified storage for improved keeping quality of cut flowers. Proc 12th Int Congr Refrig 3:339

37. Rogers MN (1980) Snapdragons. In: Larson RA (ed) Interoduction to Floriculture. Academic Press, London 107
38. Sanderson KC (1975) Harvesting, grading and quality, packing and shipping. Flor Rev 156(4040):81
39. Tees HJ, Sheehan TJ (1957) Chemical modification of geotropic bending in snapdragon. Proc Fla State Hortic Soc 70:391
40. Laurie A, Poesch GH (1932) Photoperiodism the value of supplementary illumination and reduction of light on flowering plants in the greenhouse. Ohio Agric Expt Stn Res Bull 512:8, 31, 35
41. Roberts RH, Struckmyer BE (1939) Further studies on the effects of temperature and other environmental factors upon the photoperiodic response of plants. J Agril Res 59:669
42. Haney WJ (1953) Day length manipulation to time snapdragons. Nat Snapdragon Soc Bull 2:1
43. Maginnes EA, Langhans RW (1967) Photoperiod and flowering of snapdragons. N Y State Flower Grow Bull 261:1
44. Maginnes EA, Langhans RW (1976) Flashing light affects the flowering of snapdragons. N Y State Flower Grow Bull 261:1
45. Flint H (1960) Relative effects of light duration and intensity on growth and flowering of winter snapdragon (*Antirrhinum majus* L.) Proc Am Soc Hortic Sci 75:769
46. Flint H, Anderson RC (1959) Effects of supplementary illumination on the growth and time of flowering of snapdragon. (*Antirrhinum majus* L.). Proc Am Soc Hortic Sci 73:479
47. Miller RO (1958) 50° for snapdragons. N Y State Flower Grow Bull 145:1
48. Miller RO (1959) What temperature for greenhouse snapdragons. Natl snapdragon Soc Bull 10:3
49. Miller PO (1960) Growth and flowering of snapdragons as affected by night temperatures adjusted in relation to light intensity. Proc Am Soc Hortic Sci 75:761
50. Miller PO (1962) Variations in optimum temperatures of snapdragons, depending on plant size. Proc Am Soc Hortic Sci 81:535
51. Tayama HK, Miller RO (1965) Relationship of plant age and net assimilation rate to optimum growing temperature of the snapdragon. Proc Am Soc Hortic Sci 86:672
52. Fischer CW Jr (1950) Production of a toxic volatile by flowering stems of common snapdragons and calceolaria. Proc Am Soc Hortic Sci 55:447
53. Williamson CE (1963) Plant disease affects keeping quality. In: Rogers MN (ed) The living flowers that last. Univ Mo Columbia, 19
54. Laurie A, Wagner A (1940) Deficiency symptoms of greenhouse flower crops. Ohio Agric Expt Stn Res Bull 611:19
55. Oertli JJ (1970a) Nutrient disorders in snapdragons. Flor Rev 146(3773):20
56. Oertli JJ (1970b) Phosphorus deficiency. Flor Rev 146(3774):28
57. Oertli JJ (1970c) Potassium deficiency. Flor Rev 146(3775):29
58. Oertli JJ (1970d) Calcium deficiency. Flor Rev 146(3776):51
59. Oertli JJ (1970e) Magnesium deficiency. Flor Rev 146(3777):23
60. Oertli JJ (1970f) Sulfur deficiency. Flor Rev 146(3778):28
61. Oertli JJ (1970g) Born deficiency. Flor Rev 146(3779):65
62. Oertli JJ (1970h) Iron deficiency. Flor Rev 146(3780):24
63. Mastalerz JW (1957) Boron deficiency in snapdragons. Pen State Flower Grow Bull 75:3
64. Staby GL, Robertson JL, Kiplinger DC, Conover CA (1978) Chain of life. Ohio Flor Assoc Columbus (USA)
65. Adams DG, Urdahl WA (1972) Snapdragon stem tip breakage as related to stem lignification and flower color. J Am Soc Hortic Sci 97:474
66. Mastalerz JW (1953) Low temperature conditioning of snaps. Natl Snapdragon Soc Bull 3:1
67. Bonnet JL, Smith Jr JE (1955) A geotropic response of snapdragons to treatment with trichloro phenoxypropionic acid. Natl Snapdragon Soc Bull 4:2
68. Joiner JN, Sheehan TJ, Mitchell KF (1974) Control of a geotropic response in snapdragon flower spikes. Proc Fla Sta Hortic Soc 86:374
69. Burg SP (1973) Hypobaric storage of cut flowers. HortScience 8:202

70. Larson RA, Scholes JF (1966) Effects of 8-hydroxyquinoline citrate, N-diamethyl amino succinamic acid and sucrose on vase life and spike characteristics of cut snapdragons. Proc Am Soc Hortic Sci 89:694
71. Raulston JC, Marousky FJ (1970) Enhancement of snapdragon floret color with light and floral preservatives. HortScience 5:355
72. Raulstorn JL, Marousky FJ (1971) Effects of 8–10 days of 5 °C storage and floral preservatives on snapdragon cut flowers. Fla Flower Grow 8(2):4
73. Johnson CR (1972) Effectiveness of floral preservatives in increasing the vase life of snapdragons. Flor Rev 149(3868):47, 95
74. Wang CY, Baker JE, Hardenburg RE, Lieberman M (1977) Effects of two analogs of rhizobitoxine and sodium benzoate on senescence of snapdragons. J Am Soc Hortic Sci 102:517
75. Glass JJ (1979) A study on *Freesia* returns. Vakbl Bloemisterij 34(3):42
76. Mastarlerz JW (1977) Greenhouse Environment. Wiley USA, 320
77. Barendse LVJ (1979) What measures will improve the keeping quality of *Freesia* Vakbl Bloemisterij 34(1b):28
78. Penning BV, Intransil BV (1978) A bunching machine for Freesia. Vakbl Bloemestery 33(10):14
79. Barendse LVJ (1978) The keeping quality of asparagus fern can be good. Vakbl Bloestermisterij 33(48):25
80. Carow B (1978) Keeping quality of sword ferns: different reactions to cut flower preservatives, problem with mixed bouquet. Gurtnerbose and Gurtenwelf 78:1217–1218
81. Bailey LH (1976) Hortus Third. McMillan New York

Chapter 10 Potted Plants

10.1 Chrysanthemum

10.1.1 Introduction

The demand for potted chrysanthemums has increased tremendously in the last few years, primarily because of its suitability as a potted plant. It is undoubtedly a number one potted flowering plant [1, 2, 15]. The dollar returns per sq. foot basis are highest for chrysanthemum with $ 4.12 per sq. ft. as compared $ 3.72 for lilies and $ 2.16 for poinsettia [2]. It is relatively easy to produce attractive potted mums and sell them for profit. As an indoor plant, chrysanthemums keep exceptionally well for a longer time, which makes them indispensible potted plants in the home. Futhermore, with a wide range of flower colours to choose from, it meets the liking of the vast majority of the public [3].

Chrysanthemums are generally produced in 10-, 15- or 20-cm containers. The overall form or shape of the plant is the major factor affecting the quality and marketing of pot mums. The height, spread, and the number of branches and flowers determine the overall shape of the plant. An ideal chrysanthemum plant in pot must form a circular outline when viewed from the top, with the flowers uniformly distributed over the upper surface of the plant. A slightly flat to rounded top, when viewed from the sides, is considered to be the most desirable quality of an ideal potted mum plant.

The stems of each plant in the pot should be uniformly distributed or spaced and of equal length. At the same time, these stems should be strong enough to hold the weight of the flower without bending or becoming crooked. The foliage should be lush green, uniformly distributed on the stem, and of the size representative of the cultivar. The foliage and flowers should be free from symptoms of pest and disease damage or chemical residues. The flowers must be in a uniform stage of development, but at the same time there should be proper balance between mature and immature flowers, as this is a good indication of postharvest-lasting quality of this potted plant. It is important to have the blooms very fresh and healthy when they reach the market place.

The best-quality potted chrysanthemum must have the best display value when it reaches the retail shop, have the form that is difficult to see through, and should have proper balance between flower colour and green foliage, as well as between mature and immature flowers. Most preferred flower colours in chrysanthemums are white and golden yellow, but the other types such as red, bronze, orange and pink are also sold to a certain extent.

10.1.2 Physiology and Biochemistry

The physiology of growth and flowering and senescence of flower have been described in detail in Chapter 5.

10.1.3 Postproduction Losses in Potted Chrysanthemum

Most of the post-greenhouse problems in potted mums can be traced back to the conditions under which these plants are grown in the production area. The quality of the finished product is an end result of all the cultural operations carried out during the growing period. In addition to the overall form or condition of the shoot and flowers, the conditions of the root system in the finished product can give a clear indication of the type of growing conditions and postharvest performance of the plants under indoor environment.

10.1.4 Loss Reduction Biotechnology

Postproduction losses in potted chrysanthemums can be drastically reduced by producing the plants under optimum growing conditions, adopting appropriate cultural practices, timely control of diseases and pests, as well as educating the retailers and the consumers on handling of potted mums in indoor environment. It is very important for every person in the market channel (grower – wholesaler – retailer) to understand his responsibilities in delivering a quality product to the consumer. Additional investigations are needed to standardize the techniques for handling of finished products from the time they leave the greenhouse until they reach the consumer. This will benefit the consumer in obtaining a quality product at low cost and the grower with repeated sales.

Improved loss-reduction biotechnologies in potted chrysanthemum production can be broadly grouped into:

1. Cultural practices.
2. Packing and shipping.
3. Low temperature storage of potted chrysanthemum.
4. Modified atmospheric and low pressure storage.
5. Use of chemical growth regulators.
6. Integrated pest management.
7. Care of finished product.
8. Educating the consumers and retailers.

10.1.4.1 Cultural Practices

The plants should be produced under optimum growing conditions, both aerial (light, temperature, carbon dioxide concentration) and soil (nutrition, watering etc.). The detailed discussion on this can be found in Chapter 5.

10.1.4.2 Packing and Shipping

The stage of flower development at the time of sending the plants to the market influences their performance both at the retail shop and in the indoor environment [7]. Buds that are too immature at the time of shipping may not open normally when placed in the indoor environment. At the same time, flowers that are too open will have shorter keeping quality. Therefore, it is necessary to make sure that the flowers are at the right stage of development and will have the best display qualities when they reach the market.

Potted chrysanthemums are generally sent to market when the flowers are about one-third to half open to ensure complete opening of the flowers even under relatively sub-optimal conditions [4, 6, 7]. Several growers market their potted mums when the flowers are half to fully open, while others send them to market as soon as flower petals begins to unfold. The earlier harvested flowers do not normally attain their usual flower size, particularly under poor light conditions [7]. Again, the stage of development of flowers for marketing depends on the market outlet and the distance of the market from the greenhouse range. For local markets, flowers in the fully open stage are more suitable, while for long-distance shipping or mass-marketing chains, plants in earlier stages are preferred.

Although there are no uniform grades for potted chrysanthemums, the acceptability of the plant is determined by its appearance. The height and width of the plant, number, colour and size of flower, stage of development of flower, and condition of foliage and flowers contribute to the overall appearance of the plant and indicate its lasting quality. Height and spread of the plants that are about two times the height and the diameter of the pot, respectively, are considered to be an ideal dimension for potted chrysanthemum [3]. In European markets, especially in some parts of Britain, the accepted height for a plant grown in 14-cm container is about 30 to 36 cm, with a flower head diameter of 30 cm or more [13].

The plants are usually covered with paper or plastic sleeves to protect them from any mechanical injuries due to friction. These plants are then placed in standard cardboard-shipping boxes. Usually six 15-cm pots or twelve 10-cm pots are placed in upright position in each box and are sold individually. Plants with different grades should not be mixed together in a box. In the retail shop the pot is covered with red, green or white aluminium foil to improve the appearance of the potted plant.

Because of the high cost of shipping due to the weight of growing medium, the plants are not normally shipped beyond about 100 miles. However, in recent years, some progressive growers operate their own truck lines to ship the plants over a long distance. In such cases the plants are wrapped individually in paper or plastic sleeves and loaded into the truck in trays or in racks. Trucks used for this purpose should have refrigeration and heating equipment to maintain the temperature. Potted chrysanthemums can hold up well in shipment unless they are handled very roughly.

10.1.4.3 Low Temperature Storage

Potted chrysanthemums can withstand temperatures up to 15.4 °C, but the longevity of the flower is drastically reduced at higher temperatures (19–20 °C). Low temperatures extend the storageability of pot mums. Under normal conditions out mums can be easily stored up to 2 weeks at 4 °C [5]. Before the plants are kept in low temperature storage, it is beneficial to remove the field heat by rapid precooling techniques. Staby et al. [5] compared different precooling methods, viz., hydrocooling, forced-air cooling, and hydrocooling plus forced-air cooling. They found that hydrocooling and/or forced-air cooling reduced the 7/8 mass mean cooling time (period required to lower the mean temperature 7/8 of the temperature differential between harvest and desired temperature) as compared to the conventional method (placing the sleeved box at 13 °C for 1 day followed by 1 day at 7 °C and thereafter at 2 °C). By exposing 1/6 or 5/6 surfaces of the box to free-circulating air, the 7/8 mean cooling time was decreased by 57% and 75%, respectively, indicating the importance of proper stacking of boxes in the precooling chamber (Table 10.1).

Plant longevity declines as the duration of storage at cold temperature increases beyond a certain limit [5]. The plants stored at room temperature (18 to 20 °C) had the lowest keeping quality as compared to forced-air, hydrocooling or standard cooling methods. The cultivars differ in their response to low temperature storage and pre-cooling techniques. Hydrocooling and the standard commercial method of cooling resulted in plants with lasting qualities similar in all the cultivars tested except for Puritan, where hydrocooling reduced the keeping quality of plants after 10 days storage at 2 °C [5] (Table 10.2).

Reduction in the quality of plants of cold-sensitive cultivars was exhibited through browning and/or wilting of florets, and yellowing of leaves [5]. White flowers have been shown to be more susceptible to low temperature browning of florets, while some of the other types are known to exhibit increased yellowing of leaves. These changes are permanent and render the product unsaleable.

Potted mums can be stored for about 10–15 days at 4 °C without any appreciable reduction in flower quality. The plants stored in light perform better than those held in dark. A minimum light intensity of 50 ft. candles is recom-

Table 10.1. Mass mean cooling time as influenced by number of sides exposed [5]

Surfaces exposed[b]	7/8 cooling time (h)[a]			
	Stem	Medium	Air	Mass mean
0	40	50	45	49.0
1	21	21	21	21.0
5	3	11	7	10.2

[a] Six 15.5-cm paper sleeved pots/box. Boxes were stacked 3×3×3 = total of 27.
[b] Number of surfaces exposed (0/6 or 1/6 or 5/6 moving air).

Table 10.2. Effects of type and duration of cooling on postharvest qualities of potted chrysanthemums [5]

Treatment		Frequency (%)					
Storage time (days)	Cooling	Yellow leaves	Brown flowers	Wilted flowers	Diseased plants	Brown + wilted flowers	Yellow leaves + wilted flowers
Paragon							
0	None	0	50	33	0	17	0
0	Standard	0	100	0	0	0	0
0	Hydro.	0	87	0	0	17	0
	Mean	0	79	11	0	11	0
5	None	0	33	0	17	50	0
5	Standard	0	50	17	0	33	0
5	Hydro.	0	67	0	0	33	0
	Mean	0	52	6	6	39	0
10	None	0	0	0	100	0	0
10	Standard	0	100	0	0	0	0
10	Hydro.	0	83	17	0	0	0
	Mean	0	61	6	33	0	0
	Grand mean	0	64	7	13	17	0
Puritan							
0	None	17	0	33	0	0	50
0	Standard	67	0	0	0	0	33
0	Hydro.	0	0	33	0	0	67
	Mean	28	0	22	0	0	50
5	None	67	0	0	17	0	17
5	Standard	17	0	0	0	0	83
5	Hydro.	33	0	17	0	0	50
	Mean	39	0	6	6	0	50
10	None	100	0	0	0	0	0
10	Standard	67	0	67	0	0	33
10	Hydro.	67	0	33	0	0	0
	Mean	56	0	33	0	0	11
	Grand mean	41	0	20	2	0	37

mended to retain the green colour of the foliage, but several growers use up to 400 ft. candles of light with beneficial effects [1, 4].

10.1.4.4 Modified Atmospheric Storage

Modified atmospheric storage conditions created by reducing the air pressure to 40 mmHg can extend the storage life of potted mums by nearly four to five times that of normal refrigerated storage. Several cultivars have been tested for their response to low pressure storage [8–11, 14]. The data indicate that cultivars like Neptune, Bright Golden Anne and Delware could be stored for up to 60 days, but the post-storage shelf life was drastically reduced after 28 days of storage. Therefore, the general recommendation would be to store potted mums under low pressure for less than 4 weeks.

There have been some investigations which indicate the beneficial effects of hypobaric storage of unrooted cuttings of some flower crops. In chrysanthemums, storing the unrooted cuttings at 50 mmHg pressure with 6 °C improved rooting after storage reduced the occurrence of blind shoots [11].

10.1.4.5 Use of Growth Regulators

Control of height is an absolute must in a successful pot mum production programme. Excessive stretching of stems can be checked by proper manipulation of growing conditions (photoperiod, temperature, spacing and nutrition) and use of chemical growth regulators. The former method is not very effective, although providing the plant with optimum growing conditions is absolutely necessary. During summer, when the natural day lengths are longer, the problem of excessive internode elongation is more pronounced and the use of chemical growth retardants becomes a necessity for achieving the desired effects. Furthermore, growth retardants, if properly used, improve the plant shape by promoting the growth of side shoots in relation to their position on the stem.

10.1.4.6 Integrated Pest Management

Year-round production of chrysanthemums demands a high degree of hygiene by timely control of pests and diseases. This makes it necessary to have routine pest control programmes as compared to controlling the diseases or pests after they have appeared. These programmes will have to be regular and include only the chemicals which are effective in preventing common pests and diseases from occurring in chrysanthemums. At the same time, the chemicals included should not be phytotoxic to the host plant and should not cause health hazards in the indoor environment. An integrated approach for the control of pests and disease is more desirable [12].

10.1.4.7 Care of Potted Chrysanthemums in Retail Shop and Indoors

Educating the retailers and consumers on the care and handling of potted chrysanthemum plants can go a long way in improving consumer satisfaction and in promoting repeated sales of florist products. Care tags containing brief information on care and handling of potted mums can be attached to every pot sold to enable the consumer to practice them on his own.

Potted chrysanthemum plants in the indoor environment have lower requirement for watering and nutrition. Therefore, high nitrogen levels at the time of finishing the crop will reduce the ultimate lasting quality of potted chrysanthemum. Similarly, building up of salt, too frequent watering or excessive drying of the medium in the house produces undesirable effects on the keeping quality of the plants.

Light is another major factor influencing the lasting quality of potted plants in the home environment. Potted chrysanthemums need bright indirect light or bright sunlight for at least one-half of the day to maintain the quality of flowers and foliage [4]. A good production programme for potted chrysanthemum, therefore, must involve a period of acclimazation of plants before they leave the greenhouse. This should include acclimatization for low light, less frequent watering, leaching of excess salts from the potting medium and lowering the growing temperature several degrees lower than that at which the plants were grown.

Crater [1] listed eight care and handling steps to maintain the quality of potted mums in the retail shop.

1. Unpack the box and remove the plants from the box immediately on arrival and water it thoroughly.
2. Remove the damaged leaves and blooms.
3. Display the plant at the proper place with proper arrangements for light etc. to show the best qualities.
4. Do not overcrowd the plants in the display area.
5. Water the plants as they need it, do not keep them too wet or too dry. Allow medium dryout between waterings.
6. Keep the plants out of strong and direct sunlight.
7. Do not set the plants in front of an air conditioner.
8. The display area should always be kept tidy and neat.

10.2 Poinsettia

10.2.1 Introduction

The poinsettia was named to honour the first United States Ambassador to Mexico, Mr. J. R. Poinsett, who introduced this plant into the USA from Mexico in the early 19th century. In the beginning they were grown as a garden plant for their bright red colour bracts. As a result of hard work and the outstanding leadership of the Ecke family in California (USA) and others, the greenhouse production of poinsettia rose from an almost non-existent trade to the present-day production of over 25.7 million pots sold in the United States every year. At the same time, the wholesale value increased from a mere $ 4.8 million in 1949 to about $ 79.9 million in 1981 [2, 18]. The poinsettia trade has also witnessed an increase in consumer preference, which has been unusual and unique for this crop. In terms of number of pots sold, it ranked next only to potted mums in 1981.

Postharvest Quality Requirement

Important quality considerations in poinsettia include growth habit, forcing quality of the cultivar, colour, size, and lasting quality of bracts, size and retention of cyathia, transportability and resistance to diseases and pests. An ideal potted poinsettia should have compact growth with relatively strong stems and free branching leaves that are free from injuries caused by diseases, pests, chemical sprays or other disorders, and bracts of desirable colour and the size representative of the cultivar. The bracts should be fresh, free from injuries, and hold on to the plants for a long time. The cyathia should not be large in relation to the plant and should be retained on the plant. The presence of blue colour in the bract is considered an undesirable quality in poinsettias. The responsiveness of the cultivar to altered forcing conditions makes them versatile and preferred by the growers [16, 17].

Although there are no established bases for grading, the plants are priced according to the size of the container, number of blooms per pot, size, colour and conditions of bracts and overall appearance or form of plants. Like pot mums, poinsettia have height restrictions. A top grade potted poinsettia plant must have strong stems of uniform height, which is not more than 2 to 2.5 times the size of the container [4].

Commercial Cultivars

The important varieties in the trade include Annette Hegg (A. H.) Dark Red, A. H. Diva, A. H. Brilliant Diamond, A. H. Lady, A. H. Top White, A. H. Hot Pink, A. H. Marble, V-10, V-14, C-1, Mikkel mini series, Redford Frost Noel etc.

10.2.2 Physiology and Biochemistry

Growth and Development

Flowering in poinsettia is controlled mainly by day length. Under long-days the plants remain vegetative, while short days promote flowering [19]. In nature, flower bud initiation in poinsettias takes place around September 28 [17]. Cultivar differences in response to photoperiod are well recognized and, depending on the cultivar, the plants flower in late November or early December. Most of the commercial poinsettia cultivars belong to the 9- to 10-week response group. Cultivars which flower too early need long day treatment to regulate flowering for Christmas [19, 20].

In commercial production, temperature management plays a major role in controlling the quality of the plant. The ideal temperature for promoting vegetative growth is 19.25 °C (night), which should be maintained until September 28. Flower bud initiation is poinsettia in favoured by relatively milder temperatures, around 16.50 °C. The temperature should be maintained until October 10 in order to ensure complete flower induction. For proper bract development, on the other hand, high temperatures around 19.25 °C are needed. The plants should be maintained at this temperature until pollens appear on the flower, after which it is gradually reduced to 14.3 °C for intensifying bract colour. The temperature must be changed gradually (1.2 °C in 3 days) to avoid any damage to the flower buds. High temperatures during later stages of development cause "rabbit tracks", while humid cool conditions results in development of "crud" [17].

Table 10.3. Ethylene production in stressed and unstressed poinsettias [21]

Cultivar	Ethylene production (ppb)		
	Unstressed	Stressed	Ratio of stressed/unstressed
Annette Hegg White	31	177	5.7
Annette Hegg Pink	28	143	5.2
V-14	24	31	1.3
Eckespoint C-1	24	65	2.7

Senescence of Potted Poinsettias

Most poinsettia cultivars are susceptible to leaf epinasty (dropping of leaves and bract), leaf and bract abscission and cyathia drop [20, 21, 32]. These symptoms are developed in response to ethylene synthesized due to mechanical stress created by bending of petioles of leaves and bracts. The bracts shrivel and dry at later stages, while the cyathia drop much earlier in the senescence process. Cultivars differ in ethylene production (Table 10.3). When the plants are stored at lower temperatures, the bracts develop a blue colour [4].

10.2.3 Postharvest Losses in Potted Poinsettias

Growing conditions and handling of plants at the flowering stage determine the quality of the finished products [31, 33, 35, 38]. The most important factors that are associated with poor quality include excessive stretching of the plant, splitting of the shoot apex, reduction in bract size, blueing of bract, physiological disorders and attack of pests and diseases.

Mechanical injuries result from bruising and tearing of bracts in transit, leading to discolouration or wilting. Mechanical stress caused by bending or folding of leaf lamina and/or leaf petioles and bracts stimulates ethylene biosynthesis in those tissues. The varietal differences to mechanical stress have been recognized by many researchers [21, 22]. The sleeving operation contributes more significantly to this problem than any others [28, 29]. Splitting or split top (bypass bud) is a major problem with some of the varieties like Paul Mikkelson in which the flower bud starts to initiate, but does not develop into a normal flower [20]. This removes the apical dominance, as a result of which the axillary buds start to grow and produce imperfect (partially developed) flowers. The problem is quite often observed when certain varieties are propagated early and grown without pinching.

When poinsettias are grown under still, humid and cool conditions, the cells burst, due to higher moisture content caused by imbalances between water absorption and transpiration. This is followed by oozing of latex, which when present on bracts and other tissues reduces the quality of the plant. This problem is often very severe when the growing temperature is suddenly reduced, particularly at high plant-moisture content. This phenomenon is closely related to guttation.

Another problem in poinsettia production is the "rabbit tracks", the causal factors of which are not yet known. It is often severe under high growing temperatures.

10.2.4 Loss Reduction Biotechnology

The introduction of the first long-lasting cultivar (Paul Mikkelson) paved the way for the development of postharvest handling techniques for this crop. At present, there are many poinsettia cultivars that last several weeks.

Cultural Practices

It is necessary to force the plants under ideal conditions. When other growing conditions are favourable, flowering of poinsettia takes place under short photoperiods. Controlled short photoperiods are always beneficial, as they promote uniform flowering, reduce the size of cyathia, prevent development of chlorophyll in bracts and premature flower drop, while long days usually produce clubby plants. Short days enable the growers to time the crop more efficiently.

Poinsettia cultivars vary in their temperature requirements for optimum growth and flowering. Best flower initiation occurs at 16.5 °C at night, slightly warmer temperatures (19.3 °C) are required until the end of September for promoting adequate vegetative growth. Once the flower initiation is completed, higher temperatures (19.3 °C) favour bract development. The plant should be maintained at this temperature until pollens appear in the cyathia, at which time temperature should be reduced to 14.3° to 15.4 °C for intensifying the bract colour. To produce a finished product quality in poinsettia it is necessary to regulate the temperature according to the development stage of the plant.

The fertilizer requirement of poinsettia plants can be met through the use of soluble fertilizer with irrigation water and top dressing with slow-release fertilizers [20].

As with other potted plants, close spacing affects the finished product quality by promoting excessive stretching of stems. Staby et al. [5] suggested spacing of 1 to $1\frac{1}{4}$ sq. ft. per pot of 15 cm with one pinched plant. The final spacing will have to be determined by container size, number of cuttings per pot and method of growing.

Packing and Transportation

Most of the present-day poinsettia cultivars are not suitable for long-distance transport. When the plants are sleeved for more than 48 h, the bracts and leaves do not fully recover from epinasty caused by stress ethylene [5]. The cultivars like Eckespoint C-1 and Gutbier V-14 are relatively less susceptible to epinasty than the Annette Hegg group. The bracts are also vulnerable to mechanical injuries due to friction or scratching during transport.

For short-distance shipping, the plants are wrapped in paper or plastic sleeves and then placed in cardboard boxes specially made for poinsettias. The stage of development at the time of shipping influences the final quality of potted poinsettias. In general, the plants should not leave the production area until the bracts are fully developed. Immature bracts fail to develop to normal size and have intense colour under poor light conditions.

Low Temperature Storage

Poinsettia plants require ample light and a temperature around 10.0° to 12.8 °C during storage. Under dark storage conditions the bracts become twisted and the leaves turn yellow. These conditions, if prolonged, may cause severe leaf and bract drop of drastic reduction in plant quality [5]. Similarly, temperatures below 10 °C during storage cause excessive leaf drop on removal to warm temperatures and render the product unsaleable. The sleeved plants can be stored under ideal temperature and light intensity for about 48 h without any adverse effects on plant quality [17].

Modified Atmospheric Storage

Staby et al. [30] suggested that storing sleeved poinsettia plants in 4% oxygen or 150 mmHg did not improve the storageability of the plants. Plants stored in low pressure storage had the highest endogenous ethylene upon removal from storage. Storage of unsleeved plants under reduced oxygen (4%) or pressure (150 mmHg) enhanced epinasty and endogenous ethylene levels. Hence, these methods are not of much benefit to poinsettia growers.

Use of Growth Regulators and Chemicals

As indicated earlier, use of growth retardants is an absolute necessity in successful poinsettia production. The grower has to decide the concentration and the number of sprays of growth regulators to be applied, depending on the weather conditions, the cultivar and need. Cycocel and Ancymidal (A-Rest) are the most common growth regulators used in poinsettias.

Epinasty during storage and post-storage stages can be reduced by preshipping sprays of silver nitrate or aminoethoxy vinyl glycine (AVG) [21, 22, 24, 26, 36]. However, it is important to note that they are not registered for use on poinsettias and may not be economical under all situations.

Control of Pests and Diseases

Proper care must be taken to prevent the attack of insect pests and diseases. Maintaining cleanliness in the production and storage area will reduce the occurrence of diseases. When the chemicals are to be used at a later stage when the bracts show colour, proper precautions are necessary to avoid any discolouration or bract injuries.

Consumer Education on Care and Handling of Finished Product

A plant properly handled in the home will have long-keeping quality. The importance of proper handling of the finished product should be emphasized by providing care tags with every plant sold to the consumer. In recent years, there have been several educational programmes on radio, television and newspapers on proper handling of poinsettia plants in the home. These programmes help to increase consumer satisfaction and promote goodwill with consumers.

10.3 African Violets

10.3.1 Introduction

African violet (*Saintpaulia ionantha*, family Gesneriaceae) is named in honour of its discoverer, Mr. Baron Walter van Saint Paul. It is one of the important indoor plants in the United States. The flowers are usually blue, blue-violet, bicoloured or white in colour. The plants can be induced to flower continuously in the home by regulating growing conditions. The present-day African violets are improved types of the original species developed through hybridization in the United States and Europe. At present, newer types such as miniature, trailing and long-lasting are in great demand.

African violets can produce flowers all the year round. Although they are sold as pot plants throughout the year, the demand for this plant is greatest around Valentine's Day, Easter, and Mother's Day.

Single-crown African violets in 10-cm pots are most preferred by retailers and consumers. An ideal potted African violet plant must have its leaves arranged almost horizontally, overlapping each other to cover all the noticable gaps in the pot, which would otherwise detract from the appearance of the plant. The foliage should be free from cold water spots, diseases and pests. The flowers should be fresh, representative of the cultivar, and stand well above the foliage. An ideal plant should have great number of flowers in a cluster at the centre. Although there are no established standards or grades, the plant must have a good display of flower with at least 5–6 cymes and 10 to 12 fully opened flowers at the time of sale. Since the leaves and petioles are susceptible to breaking, proper precautions in transit and handling in production area and retail shop are necessary to deliver quality products to the consumer.

Most of the cultivars that are grown today were developed from then named cultivars introduced during the mid 1930's. Some of the newer types include Ballet, Rhapsody, and the Optimara series. In addition to these, some of the old varieties like Neptune and Blue Boy are still grown to a certain extent. At present, the efforts of breeders are concentrated to evolve long-lasting African violets as well as miniature and trailing types [39].

10.3.2 Physiology and Biochemistry

Flowering in African violets is regulated by growing temperature and light conditions. The response of African violets to temperature is unique, in that the growth of plants is greatest at relatively higher night temperature and lower day temperatures (10 °C) [39, 40]. The optimum night temperature for most rapid growth and flowering is 21 °C [39]. Went [41] demonstrated that African violets grow best and flower profusely when night temperatures are maintained at 20 to 23 °C with day temperature of 14 °C.

The temperature of the water used for irrigation plays an important role in determining the quality of the plants. Cold water drops cause the development of white to yellowish brown spots, rings or streaks at the point of contact on leaves [39, 40]. This problem occurs when the difference between leaf and water temperature is more than 16.5 °C. Elliot [40] suggested that cold water moving through the leaf cells that are warmer than water results in collapse of the cell wall and destruction of chlorophyll, with the result that the cell turns white or yellowish brown. These symptoms are irreversible and permanently detract from the appearance of the plant.

Radiant flux density is most critical in African violet production. The optimum light intensity for rapid growth and profuse flowering has been shown to be 10.8 to 11.8 klx (1000 to 1150 ft. c.) [42, 45]. Solar radiant flux density above about 13.8 klx (1300 ft. c.) causes destruction of chlorophyll, browning of leaves, reduction in growth or stunting. In contrast, lower light intensities below 3.24 klx (300 ft. c.) reduce flowering in African violets [45].

Kiplinger [43] suggested that African violets grow and produce better-quality flowers under artificial light than in natural daylight. The time required to produce 10 cm potted African violets can also be reduced from 4 to about 2 months by artificial lighting. The optimum light intensity for artificial lighting has been found to be 6.5 klx for 12 to 18 h per day. The irradiated plants produced heavier dark green foliage and early flowering [46]. Zoerb [47] suggested a standard cool-white fluorescent lamp, whereas Boodley and Horton [48] recommended wide-spectrum Grow-Lux fluorescent lamps for successful production of quality African violets.

10.3.3 Postharvest Losses in African Violets

The majority of the post-production losses can be attributed primarily to faulty growing conditions, or improper handling of finished products. Reduction in leaf size or colour, presence of water spots or ring spots on the foliage, reduction in number, size, and quality of flowers, shorter duration of flowering, presence of diseases or insect pests and flower or foliage damage caused by rough handling affect the final quality and marketability of potted African violets.

10.3.4 Loss Reduction Biotechnology

Loss reduction measures must begin in the greenhouse right from the start of the crop on the propagation bench and followed throughout the production phase to harvesting, packaging, shipping, retail selling and the consumer. These would include cultural (light, temperature and nutrition), packaging, shipping, and handling of finished products at the retail shop and by the consumer.

1. Growing of plants under optimum temperature and light intensity (preferably under artificial light) is important in order to produce the best quality plants in the time for the holiday.
2. The potting medium for African violets should have an adequate amount of organic matter to keep it moist all the time.
3. Excessive drying of potting medium can cause root damage and affect subsequent development of the plant.
4. The method of watering is very important to avoid cold water leaf spotting. As the leaf spotting is produced when the temperature of water used for watering is 16.5 °C less than the leaf temperature, the water temperature should be similar to that of the leaf tissue (21.4 °C). The damage can be prevented by using spaghetti tube or capillary watering techniques.
5. Nutrition of plants in the production area is another important consideration in African violet production. African violets are sensitive to higher soluble salts. Therefore, the fertilizers for African violets should be low enough to avoid building up of salts in the growth medium. Shanks [44] recommended 75 ppm each of nitrogen and potassium at every irrigation.
6. Normally, the plants will be ready to go to the retail shop when they have a good display of flower (with at least 10 to 12 flowers fully open).
7. Most growers ship their potted African violets in 7.5- or 10-cm pots. The plants are usually covered with paper or cellophane sleeves and then packed in the cardboard boxes. The plants in the box are separated from adjoining ones by placing cardboard dividers or plates to avoid damage caused by friction or displacement of plants inside the box.
8. The plants are shipped by air or refrigerated trucks.
9. On arrival of the consignment at the retail shop, it is very important to immediately unpack the plants, remove the sleeves and water.
10. The plants do not keep well in darkness. The flowers can be kept in good condition for a longer period by placing the plants under artificial light. Therefore, precaution should be taken to avoid darkness in the display area.
11. The temperature should not at any time be below 12.3 °C.
12. The soil should be kept moist at all times.
13. Each plant coming out of the production area must have a care tag to facilitate the consumer in extending the aesthetic life of potted African violets.
14. Proper plant protection measures are necessary to prevent pests and diseases at all stages.

10.4 Begonia

10.4.1 Introduction

Begonias are considered as one of the most versatile ornamental plants grown commercially by florists. They form an attractive potted flowering plant with a wide variation in flower colour, form, size and texture. Some of the species of begonias have beautiful markings on the leaves, making them attractive foliage-potted plants. As bedding plants, begonias are more popular than some of the annuals and are very well suited for growing on flower stand or as specimen plants.

The genus *Begonia*, which is named after Michael Begon, a botanist from Santa Domingo, consists of nearly 1000 species, of which approximately 200 are under cultivation in different parts of the world. Botanically, begonias are classified into rhizomatous, fibrous rooted, or tuberous begonias. The most important species from the horticulturist's point of view include the Rieger begonia (*B. hiemalis*), fibrous rooted begonia (*B. samperflorens*), rhizomatous begonia (*B. tuberhybrida*), and Christmas begonia (*B. socotrana*) [49].

Like other potted flowering plants, the overall appearance of the plant is the major factor which determines the quality in begonias. The stage of plant development at the time of selling is also very important and the grower should make sure that the plant will be in best possible display when it reaches the retail outlet. Generally, the plants will be ready to be sent to market when at least one-half of the total number of flowers are fully open and look fresh.

In some of the begonia cultivars, especially the monoceous types, the male flowers abscise in the bud stage. Hanisch et al. [59–62] demonstrated that the abscission zones are formed in elongating pedicels. They further suggested that removal of flower buds accelerates pedicel abscission, while separation of the pedicel from the parent plant retards it. Application of auxin (IAA) could substitute for flower bud. Ethylene and abscisic acid (to a limited extent) were found to promote flower abscission [60]. These hormones do not interact with each other, but absisic acid was found to act additively with ethylene [61]. The IAA level in the abscissing flower buds was reduced to 1% of that of normal flower buds and the seasonal variation in flower buds drop in *B. davisii* coincided with changes in the IAA level of buds.

10.4.2 Post-Production Losses in Begonia

Post-greenhouse life of potted *Begonia* is determined by the ability of the plants to grow and flower profusely under indoor environment. Plants that were never acclimatized to home environment where the light intensities are relatively lower than in the production area may not perform well under such conditions. The Rieger begonias revert to the vegetative phase under long days and higher temperatures (above 21 °C). Secondly, the nutritional status, as well as the health of the plant when it leaves the greenhouse, are very important factors that influence the post-production life of potted flowering plants [58]. A poorly developed root system with brown or rotten roots is an indication of inadequate care in the production area and may directly reduce the plant's ability to continue growing and flowering at reduced light intensity and temperatures inside the home. Similarly, plants deficient in major nutrients at the later stages of the forcing schedule usually do not perform well at home unless they are supplemented with additional fertilizers. In some of the species of begonias with monoceous flowers, the staminate flowers abscise prematurely.

10.4.3 Loss Reduction Biotechnology

The greenhouse manager or grower can do a lot to check the losses that occur in post-production phases. It is very important for flower growers to understand the physiology of growth and flowering in various begonia species before even planting a crop [50–58].

All the growing conditions will have to be at optimum level in order to produce an excellent quality plant. Growers can either follow one of the standard flowering schedules developed by various workers or develop their own through experience. Acclimatization of the plants to conditions prevailing indoors is a must to prevent undesirable effects, such as leaf and flower abscission or reduction in growth and flowering.

Begonias prefer ample indirect sunlight throughout the day or full sunlight for at least half of the day. The plants need adequate support to avoid damage due to breaking of shoots. A care tag supplied with the plant is helpful in educating the customers in handling their plants and deriving maximum satisfaction out of this marvelous plant.

10.5 Geranium

10.5.1 Introduction

The present-day florist geranium (*Pelargonium hortorum* Bailey) is the hybrid derived from *P. zonale*×*P. inguinaus* and *P. domesticum*. Geraniums are used as flowering pot and bedding plants in the florist trade [66]. The primary market for geraniums is in the spring, when people in the north want something to plant outdoors. They are also used as a gift on different occasions. In the recent past the demand for 10-cm seedling geranium and cultivars with variegated foliage has increased tremendously. In 1979, approximately 46.7 million potted geraniums were sold in the United States of America [2].

An ideal potted geranium must hold the petals, continue to grow and flower profusely in the indoor environment and be free from diseases, pests and physiological disorders such as oedema. The plants will be ready for sale when there is a good display of colour on the plant. The height and the overall form of plant are further important quality attributes of a potted geranium [63–68].

10.5.2 Post-Greenhouse Losses in Geranium

Probably nothing is more detracting and dissatisfying than buying a potted flowering plant which quits flowering as soon as it is taken home. Petal shattering is one of the most serious problems that a consumer has to face. Flower shattering occurs when hormonal imbalançe is created within the plant. Certain cultivars have been shown to be more susceptible to flower shattering than others. The presence of ethylene in the atmosphere accentuates the problem of petal shattering.

Considerable losses can occur due to rough handling operations or due to the ignorance of retailers on the care and handling of potted geraniums. The new more succulent shoots are vulnerable to mechanical injuries. Excessive stretching of shoots makes the plants more susceptible to breaking.

Botrytis blight is often a major problem, both in stock plant range and forcing greenhouses. Oedema is the most common physiological disorder in geraniums. This disorder develops when humidity in the atmosphere is high and plants have surplus water. It is characterized by corky areas on the undersurface of leaves and petioles [63].

10.5.3 Loss Reduction Biotechnology

To have a longer-lasting plant at home, it is necessary for growers to produce quality plants. General production programmes have been developed both for seedling and vegetatively propagated geranium cultivars, and are very helpful to growers interested in producing high-quality potted geraniums [69–71].

Since the height of the plant will have to be in proportion to container size, proper use of growth regulator is important to produce plants of desired shapes. Several workers have recommended Chlormequat (CCC) as a foliar spray or soil drench application to reduce plant height, to hasten flowering, and to increase the number of flowers per plant [72, 73]. Ancymidol spray has also been suggested for this purpose [92].

Geraniums are one of the crops which benefit most by CO_2 enrichment in the greenhouse, especially during the winter months when ventilation is a problem. Plants grown in CO_2-enriched atmosphere grow faster and flower early. The optimum range of concentration of CO_2 was found to be between 500 and 750 ppm.

Proper care during packing and transportation is necessary to reduce petal shattering and leaf fall. Care and handling of plants by the consumer and retail florist are equally important.

10.6 Kalanchoe

10.6.1 Introduction

The *Kalanchoe*, once regarded as a minor flower crop, is becoming increasingly important due to the wide range of leaf form and colour of flowers available for selection. The flowers have excellent keeping quality under indoor environment. The present-day potted *Kalanchoe* cultivars are hybrids developed from the cross between the original *K. blossfeldiana* and *K. velutina*. In the United States, most *Kalanchoe* improvement is being done by J. T. Irwin and Jim Mikkelson. It was introduced into the florist trade in Germany in 1932 by Robert Blossfeld. Large-scale greenhouse production of *Kalanchoe* in the United States started only 6 to 7 years back [74].

Kalanchoe are generally produced in 10-, 12.5- or 15-cm pots. The overall appearance of the plant is the reflection of the type of growing condition and the handling techniques during the pre- and post-production stages. A top quality *Kalanchoe* must have at least half the total number of flowers fully open. The flowers in the bud stage will not open if conditions during transport and marketing are unfavourable. The leaves should be dark green, fresh, and free from all injuries.

There are numerous patented cultivars available for growing and many more are being added every year by the hybridizers. Most of the earlier hybrids require 3 to 4 weeks of short-day treatment compared to the newer ones which need 5 to 6 weeks of short-day treatment [74–79, 81, 82, 84–86]. Some of the cultivars are Mace, Telstar, Vulcan, Tetra Vulcan, Thor, Tom Thumb, Brillant Star, Golby Melody, Cactus Candy, Jupiter, Goldrand, Rotglut, Morgensonne and Amethyst. Of these only a few, like Mace, Golby Melody, Telstar, Cactus Candy, and Jupiter are more popular than others.

Crocked stems, damaged or pale-coloured leaves, premature fading or floret drop can greatly reduce the quality of potted *Kalanchoe.* Like other pot plants, *kalanchoes* also have height restrictions, and in general plant height should not be more than about 2.0 times the height of the container [95].

Air pollution with excess ethylene in the storage environment results in premature fading of petals, desiccation, or failure of buds to open. Plants exposed to 0.5 ml of ethylene per litre can cause serious irreversible injuries in *Kalanchoe.* Similarly, extended dark storage at 10 °C causes yellowing of lower leaves, and fading of flowers [5, 80]. Many *Kalanchoe* cultivars are susceptible to powdery mildew and crown rot diseases.

To ensure a good finished product quality, it is very important to produce the plants under uptimum light and temperature conditions [74–79, 81–82, 84–93]. Although it is a succulent plant, it needs an adequate amount of light. Most of the European cultivars require a light intensity around 27 klx from May to August. During the summer months, the plants need to be shaded to protect them from excess light intensity, while during other times full sunlight is recommended. Plants grown at reduced light will have a lower number of basal shoot and lack fullness or adequate spread [78]. Plant height and spread can be effectively controlled by the use of B-Nine [83].

Kalanchoe needs a medium which is well drained and aerated, with pH adjusted between 6 and 7. A pH lower than 5.5 causes browning of basal leaves, while a pH above 7 produces yellowing upper leaves [94].

The stage of plant for sale depends upon the distance of market and cultivar. In general, at least half the total number of flowers should be open before they are shipped. The plants should be fed before they are sold, so that the flowers hold up better. Plants developed ahead of schedule can be maintained at 10 °C up to 4 weeks to slow down bud development. Plastic or paper sleeves can be used to cover the plant to avoid mechanical injury in transit. Sleeves should be removed on arrival in the retail shop and the plants should be maintained at 16–20 °C. Since the plants are usually produced under bright light, the finished products should not be placed in a box or dark room for a long time, to prevent yellowing of leaves. The rooms where kalanchoes are stored should be completely free from ethylene, as even 0.5 ml of ethylene per litre of air causes premature fading of flower, and failure of immature flower buds to open [80].

10.7 *Easter Lily*

As the name indicates, Easter lily (*Lilium longiflorum*) is produced only for Easter. Easter lily is a highly specialized crop because the date of Easter does not remain constant every year and hence the growers have to skillfully manipulate all the cultural practices to produce the crop at the right time each year. The pre-cooled bulb takes about 90–110 days for forcing depending on temperature and light intensity.

Major quality considerations in Easter lily are final plant height, length of lower leaves, bud count, stage of flowering, opening and arrangement of flowers, freedom from diseases, pests and physiological damage, and the overall form of the plant in relation to container size. Among these factors, the height, bud count and overall appearance of the plant are probably the most important considerations for a quality plant. Bud counts of four to five are considered to be minimum for most of the commercial Easter lily cultivars [107]. A well-developed healthy root system with white root uniformly distributed in the medium is a reflection of quality in potted Easter lily, as it indicates proper care during forcing. The soluble salt level of the medium should be low enough to ensure longer indoor life of plants.

Reduction in useful life of potted lily can be brought about by senescence of flowers, yellowing and abscission of lower leaves, or failure of immature buds to open. Like other crops, senescence of flowers in lily is accompanied by irreversible wilting, shrivelling and necrosis of petals. Exposure of plants to toxic levels of ethylene can cause a similar response. Removal of anthers at early stages has been shown to extend the lasting quality of these flowers [107].

As indicated earlier, the quality of potted lilies is affected by excessive stem elongation, small or too long lower leaves, yellow lower leaves, fluoride damage, blasting, reduction in bud count, and attack of pests and diseases. These factors can be classified into four major categories; cultural, mechanical, physiological and pathological.

Insufficient care during moving, packing and transportation can result in mechanical damage such as breaking of stem, tearing or crushing of petals and leaves. When clay pots are used, rough handling can break the pot, with the result that the root system can be disturbed.

Flower bud blasting, abscission of lower leaves, leaf scorching and excessive stretching of plants are some of the undesirable qualities in potted lilies. Flower buds in lilies may stop developing and die prematurely during the greenhouse-forcing phase. Insufficient light levels, too high or low temperatures, prolonged vernalization treatment of bulbs and inadequate fertilization can cause flower bud blasting.

Yellowing and abscission of lower leaves can be due to several reasons, root injury, inadequate or excess fertilization, insufficient light and air circulation (due to crowding of plants), water stress or high fluorine levels in soils [115]. Abscission of lower leaves detracts from the appearance of the plant. Leaf scorching in lilies can be due to high fluorine levels in soil or irrigation water or high salt content in soil [114]. Incorporation of materials, like super phosphate, German plant moss, perlite or vermiculite, increases the level of fluorine in potting medium and produces undesirable effects. In addition, high soil pH can also cause severe leaf tip burning. The symptoms of leaf burning persist until the death of the leaves.

As with other potted container plants, potted lilies have height restrictions. Depending on the cultivar, lily plants over about 45 – 56 cm, including the container, are considered to be too tall. Overcrowding of plants in the greenhouse or insufficient nitrogen can produce unusually tall plants with pale yellow leaves, and poorly developed flower buds.

Pest and disease infections ruin the appearance of the plant and the overall acceptability.

Quality is an end product of efficient management of all the contributing factors, such as adequate vernalization of bulbs, forcing environment, nutrition, pest and disease control, proper handling of the finished product and its care at retail and consumer level. Unlike other cut flowers and potted plants, the demand for Easter lily is only at Easter and the plants that are not sold by that holiday have no value whatsoever. To have the plants in flower at the right time for Easter every year requires full understanding of the production technology in this crop. The complicated physiology of this crop, together with less flexibility in the time factor, means that application of improved loss reduction biotechnology has added importance in Easter lilies. The major concerns of the lily growers are to reduce blindness and flower bud blasting, produce a plant with a strong stem of the desired height having uniformly distributed dark green foliage, lower foliage that is not too narrow or long, and free from all injuries. It is absolutely necessary to time the crop in such a way that it will be ready to go to market about a week before Easter, to ensure its complete sale. The improved loss reduction biotechnologies in Easter lilies can be categorized as follows:

1. Improved cultural practices [96, 98, 99, 100 – 115].
2. Storage [96, 107].
3. Care and handling of finished product [4].

The success of lily growing primarily depends on the skill and desire of the grower to integrate all the factors during forcing. Temperature and photoperiod are the most important factors determining the quality of the final product and time of flowering. Several investigations have been conducted to devise a precise production schedule for Easter lily.

Plants in the puffy flower stage can be stored for some time without any adverse effects on the quality. Several factors affect the storageability of plants at this stage. Unfortunately, it is not very clear at present how these factors influence plant quality. Tayama et al. [107] listed some of the steps that are beneficial for storing potted Easter lilies:

1. Harvest the flowers at white puffy stage.
2. Water before storing.
3. To help prevent *Botrytis*, mist with Benlate at 0.5 g per litre (8 oz. per 100 gal water) prior to storage.
4. Use paper sleeves instead of plastic to cover the plant before they are boxed.
5. Stack the boxes in such a way to expose at least one side of each container or box to avoid undesirable warming up.
6. Whenever possible, it is better to store them under light.
7. Do not store beyond 2 weeks.
8. Optimum temperature for storage is around 1.65 °C with 92% relative humidity.
9. Upon removal, allow the plants to warm up slowly to avoid water stress due to rapid warming of aerial part as compared to root system.
10. Keep the storage atmosphere free from ethylene gas.

If properly cared for, potted lily flowers can be enjoyed for over 10 days. A plant with a well-developed and healthy root system and at least two to three unopened flowers lasts longer than plants with fully opened flowers. High soluble salts and fluorine levels in soil, dark indoor setting, high temperature, ethylene pollution, and low relative humidity affect the lasting qualities of lily plants in the indoor environment. At the same time, plants that are finished at relatively lower temperatures perform better than those grown continuously at warmer days and nights. Watering of plants in the indoor situation is very critical and is often ignored. Watering should be based on need, and care should be taken each time it is done to allow the water to drain out. Water stress or too much water often causes problems. Removing the anthers before the pollens are shed keeps the petals clean and attractive. Wilted or faded flowers should be removed to encourage proper development of unopened flowers.

10.8 Cyclamen

The cyclamen plant received its name from the Greek word kyklos (circle). In its native habitat, cyclamen grows and flowers during mild weather and goes into dormancy in summer. For years it has been an important potted flowering plant in Europe. Cyclamen is a herbaceous plant with cluster of blueish-green ovate leaves and attractive sympetalous five-parted reflexed flowers that hold well above the leaf canopy. In several cultivars the leaves and petioles have silvery markings. Both diploid and polyploid forms are available in the trade. The flowers, which are either single or double, have fringed, frilled, crested or simple petals.

An ideal potted cyclamen must have uniform flowering, an open centre with leaves growing away from the centre and a cluster of buds uniformly distributed in the centre. One flower per leaf has been found satisfactory. The foliage must be fresh, crisp and free from injuries.

Cyclamens are propagated through seeds, by which method it takes about 80 to 90 days to produce seedlings, Greenhouse forcing in cyclamen involves growing the plants at 20 °C (night) until six to seven leaves are folded, after which it is dropped to 17 to 18 °C [121]. Flower initiation in this crop does not take place until plants attain a certain height. Usually, flower bud initiation takes place in the axils of the sixth true leaf onwards [121]. Plants have been shown to have 35 or more leaves are unfolded before the flowers are visible and take about 6 weeks from the 35-leaf stage for flowering. Gibberellin (25 ppm GA_3) application has been shown to hasten flowering.

Potted cyclamens have the best keeping quality if properly watered and maintained under bright, cool (18–20 °C at night) conditions. Individual flowers last about 4 weeks to give a total post-production life of 6 weeks or more. The soil medium must always be uniformly moist but not wet. The plants should be fertilized before they leave the greenhouse.

10.9 Gloxinia

The florist's gloxinia (*Sinningia speciosa*), with its rosette of leaves growing on a short stem, forms an exceptionally attractive potted plant. The leaves are pubescent, and 15 to 28 cm long. Trumpet-shaped 5- to 10-cm-wide flowers are borne above the foliage in nodes and open in succession. The flower colours include red, white, purple and bicolour. Through hybridization and selection, double forms are also available in the trade. Gloxinias can be produced all the year round with peak production from March to November. On a commercial scale, the crop is grown from seed and it takes about 150 to 190 days to produce them in 12.5-cm pots.

Top quality gloxinia in a pot must have at least one-fourth of the total flowers open at the time of sale, with the remaining ones in various stages of development. The flowers should remain well above the foliage and be free from browning, mechanical damage and discolouration. The leaves should extend slightly below the rim of the pot and be free from cracking. The root system should be healthy and uniformly distributed throughout the root ball. Faded flowers must be regularly removed.

Three-month-old seedlings are usually used for commercial production. A soil-less mixture consisting of soil : peatmoss : perlite (1 : 1 : 1) having a pH of 6 can be used with good success. Top grade plants can be produced by forcing the plant at 21 °C (night) and, 25.8 klx light with a relative humidity of 50% to 70%. With proper fertilization and use of plant growth retardants, it is possible to produce quality plants and also to intensify flower colour [121].

Postharvest quality in gloxinia is drastically reduced by browning, fading and discolouration of flowers, yellowing and cracking of leaves, and due to diseases and pest infestation. Poor light in the postharvest environment results in browning of flower buds. The quality of potted gloxinias can be improved by maintaining them under bright indirect light (75 to 250 ft. c.) and at 21.24 °C (day) and 16 to 18 °C (night). Watering should be uniform to maintain proper moisture on the soil. Splashing of cold water on leaves should be avoided, as it produces yellow spots. Feed the plants with complete fertilizers once in 2 weeks. Under ideal conditions, individual flowers last about 4 to 6 days, while the postharvest life of the whole pot is around 2 to 4 weeks or more [121].

10.10 Calceolaria

Calceolaria herbeohybrida Voss, because of the inflated pouches in the flower, is also called the pocket book plant [116]. The flowers, 2 to 3 cm in diameter, are produced in masses on the top of medium-sized leaves, giving an overall height of 30 to 45 cm. Calceolarias are available mainly during January to May. The flower colours include yellow, red, and bronze with deeper-coloured freckles.

The florist's *Calceolaria* cultivars are grouped into grandiflora (3.75 to 5.0 cm flowers on 30 to 40 cm plants), grandiflora primula compacta (4.5 to 5.0 cm flowers on 20 to 25 cm plants), multiflora (2.5 to 4.0 cm flowers on 25 to 30 cm plants) and multiflora nana (2.5 to 3.0 cm flowers on 30 cm plants) [117].

A top grade potted calceolaria must have several clusters of flowers displayed well over the foliage on strong, stiff, and upright stems. The inflorescence should be complete without any missing florets and free from all injuries. The root system must be sound, healthy and uniformly distributed. The foliage must be turgid, and free from injuries. One plant in a 12.5- to 15-cm pot gives a good display.

Calceolarias are propagated from seeds which are minute (20000 to 40000 seeds/g). Flower initiation in calceolarias takes place at low temperatures (15 °C) [118, 119] and under long days (14 to 15 h critical day length) [4]. The plant and flower quality in this crop was found to be determined primarily by interaction between photoperiod and temperature [4]. Several forcing schedules are available for successful production of calceolarias in the greenhouse.

Too dry soil or high irradiance causes fading or scalding of flowers. Similarly, warm dark conditions promote excessive stretching of flower stalks and result in breaking of flowers. Over-watering also damages the root system, thereby drastically reducing growth and flowering. In addition, splashing of water drops on leaves causes irreversible damage by producing spots and stem rots. Ethylene in the holding environment will cause flower fading and abscission.

The quality of potted calceolaria can be improved by growing them under optimum conditions and handling them properly in the marketing channel. They should be unpacked promptly on arrival and maintained under bright indirect light (12.5 klx) and at 5 to 7 °C night temperature with 13 to 18 °C during the day. Watering should be based on need, and be uniform. Calceolarias do not need additional fertilizers while in flower. Under most favourable conditions calceolarias have a post-production life of 2 to 4 weeks.

10.11 Streptocarpus

The present-day *Streptocarpus* is a hybrid involving various species with a very short stem (almost stemless) and rosetted leaves which are 30 cm long in an asymmetric clump. The flower stalks, carrying several flowers, rise from the centre of the crown and stand well above the foliage. Individual flowers, 5 to 10 cm long, have crested or fringed petals which are white, pink, purple, lavender or blue in colour. Some of the best-known hybrids are the Nymph series, Diana, Fiona, Karen, Marie, Paula, Tina, and Halen. The crop is grown from seeds or leaf sections. It takes about 6 months to produce a quality pot. Under favourable growing conditions, *Streptocarpus* flowers all the year round.

As it has a shallow root system, azalea pots are most suitable. Soil media should be allowed to dry between watering. Growing temperature is extremely important for *Streptocarpus* culture. Most of the hybrids do well at 16.0 °C (night) and 50% relative humidity. *Streptocarpus* plants need 12.9 klx light for producing quality flowers [120].

The quality of potted streptocarpus can be improved by growing them under ideal conditions and maintaining them under 12.9 klx light, 21 to 24 °C (day) and 16 to 21 °C (night). The soil must be uniformly moist but not wet. The roots can be easily damaged by over-watering.

10.12 Cineraria (Senecio hybridus)

Senecio hybridus is believed to be the result of crossing among *S. cruentus, S. heritieri* and other species [120]. The florist's cineraria is a showy pot plant with daisy-like flowers in a cluster. The inflorescence are present over and within the foliage. Individual flowers are 2 to 5 cm in diameter and come in many shades of bright, red, purple, blue and salmon with white eyes. The present-day greenhouse cinerarias are grouped into grandiflora (early with few larger flowers) and multiflora (several small flowers) types [117]. The leaves are rounded, 5 to 7.5 cm across, and form a rosette giving an overall height and width of 30 to 40 and 25 to 40 cm, respectively.

A top grade potted cineraria must have a compact spray of flowers free from foliage or other extraneous materials sticking to their surface. About one-third to one-half of the total number of flowers must have opened at the time of sending them to market. Dry, off-coloured or faded flowers, or dry pollens on the flowers reduce the acceptability in cineraria. In general, one uniformly grown plant free from mechanical injuries, pests and diseases in a 12.5- to 15-cm pot produces the most acceptable potted cineraria.

Cineraria are propagated by seed during September to November. Greenhouse forcing in cinerarias includes growing the plants at 18 °C for 8 to

10 weeks before 6 weeks of low temperature treatment at 9 to 12 °C for flower initiation [120]. Gibberellins can be used to partially replace low temperature requirement, but this reduces quality by promoting excessive stem elongation [82]. The plants after flower initiation should be grown at 12 to 13 °C for 9 weeks for proper flower bud development. Temperature above 13 °C during this period have been shown to reduce plant quality. High fertility, particularly nitrogen, reduces the plant quality by excessive enlargement of leaves.

In general, cinerarias have short post-production life because they cannot tolerate warm dry weather. Post-production losses in cinerarias can result from under-watering, excessive fertilization, improper forcing technique, pests and diseases and unfavourable conditions in an indoor climate. Drying of the soil is the major problem faced by the consumer.

The lasting quality of cinerarias can be improved by keeping them in cool (13° to 18 °C day and 7° to 13 °C night) well-lighted situations (indirect light). Watering is most critical in the indoor environment.

10.13 Rhododendron

Azaleas (*Rhododendron obtusum*) have been grown as a flowering potted plant for many years. However, in recent years, their popularity is reduced because of the very high cost of production and longer time required to produce quality products. It is still one of the important flowering potted plants in Europe, where its culture has been perfected better than in other countries. Azaleas are usually produced for Christmas and Easter. However, year-round production is possible under controlled environment.

Azalea is a small woody evergreen shrub reaching a height and width of 30 to 60 cm and 30 to 45 cm, respectively, in pots. Flowers are single, single hose-in-hose (two layers of simple petals), semi-double, semi-double hose-in-hose, or double hose-in-hose. The flowers vary in size, shape and colour. The most common colours are white, pink, red, purple, variegated pink and white and red and white.

Top grade potted azaleas must have uniform distribution of flowers and buds and proper balance between flower colour and green foliage. In general, plants must have about one-third to one-half of the flowers open at the time of sale. A quality plant must be free from new vegetative growth, wild or long light shoots, and have most of the branching at or near the soil surface. The root system must be healthy. An individual flower must last at least 5 to 10 days and the whole pot must be in colour for at least 4 to 8 weeks. The size of the plant in relation to the container size is an important consideration in the marketing of azaleas.

Although azaleas can be produced all the year round, the major production is done during January to April. Liners are used to produce the commercial crops. Rapid and vigorous vegetative growth can be expected when temperatures are around 25° to 30 °C during the day and 16° to 18 °C at night. In addition to air temperatures, soil temperature has a profound influence on the final quality. Soil temperature of 23 °C has been shown to be optimum to produce quality potted azaleas. Quality, intensity and duration of light affect azalea growth. However, these should be considered in relation to growing temperatures. Detailed forcing schedules are available to growers.

Reduction in post-production lasting quality of azaleas results from flower and leaf abscission, fading or discolouration of flowers, physical injuries to plants and flowers, and disease and pest infestation. These losses can be minimized by following proper postharvest-handling techniques. Azaleas are graded based on crown diameter, and average number of flower buds. In general, a plant having 32 cm diameter must have 45 flower buds. However, none of these methods includes post-production lasting quality as a criterion.

Azaleas are frequently covered in paper sleeves to protect the flowers and are boxed for long distance transport. Maintenance of proper soil moisture is an important consideration in long-term shipping. On arrival in the retail shop, it is essential to unpack them from boxes and check the soil moisture. The keeping quality of these plants can be extended by keeping them at 4 to 20 °C. Bright indirect light and mild temperatures (16 to 21 °C day and 10 to 16 °C night) in the storage room improves the quality of azalea plants. The medium should never be allowed to dry out, although excess watering affects the quality. Care tags should be provided with the pots to help the consumer in maintaining the plants for longer period of time.

10.14 Hydrangea

Hydrangea has been a popular garden plant in Asia and Europe for several years. It is a small to medium deciduous shrub with glossy dark green leaves, 10 to 15 cm long. The inflorescence is a spectacular sphere with pink, white or blue flowers. Hydrangeas are produced mainly for Easter and Mother's Day.

An ideal potted *Hydrangea* must have 90% of the flowers in a cyme fully open and be in full colour at the time of sale. The stems must be strong enough to hold the flowers in upright position. One branched plant in a 10- to 15-cm pot produces the most acceptable product.

Flower intiation takes place rapidly at 13 to 18 °C under both long and short days. However, normal flower development takes place only during long days. Under short days, the flower buds enter a state of dormancy and require cold treatment to break it. The plants, after complete leaf fall, are subjected to 2° to 9 °C for 6 weeks to promote flower bud expansion. Presence of light during cold storage has been shown to help in breaking dormancy. However, gibberellic acid (100 ppm GA_3) can replace the requirement for cold treatment or long photoperiods.

Post-greenhouse life is prolonged by maintaining proper moisture in the soil, and placing them in a cool, well lighted place.

References

1. Crater GD (1980) Pot mums. In: Larson RA (ed) Introduction to Floriculture. Academic Press 261
2. Anon (1982) Floriculture crops-production area and, sales 1980 and 1981. Intentions for 1982. Statistical reporting service. US Dept Agric Wash DC
3. Laurie A, Kiplinger DC, Nelson KS (1968) Commercial Flower Forcing. McGraw-Hill, New York
4. Staby GL, Robertson JL, Kiplinger DC, Conover CA (1978) Chain of life. Ohio Flor Assoc, Columbus, USA
5. Staby GL, Thompson JF, Kofranek AM, Thompson VR, Walter VR (1981) Cooling of potted chrysanthemums. HortScience 16(4):566
6. Tayama HK, Kiplinger DC (1968) Pot chrysanthemum culture. Leaflet No 142 Co-op Extn Serv Ohio State Univ, Columbus
7. Walter VR (1974) Studies on the keeping quality of potted chrysanthemums as affected by differential degrees of flower development. Ohio Flor Assoc Bull 539:7
8. Della-Vedowa RP, Rose BA (1950) Altitude tests on flowers. Lockheed Air Corpo Prog Rep SLR-942
9. Anon (1975) Grumman Dormavac hypobaric storage and transportation of perishable commodities. Grumman Allied Ind Inc, New York
10. Byers B (1977) In: Dewey DH (ed) Grammon Dormavac system controlled atmosphere for storage and transport of perishable agricultural commodities. MSU 28:82
11. Butler RE (1976) Promise of longer storage under low pressure confirmed in UK. Grower (London) 86:395
12. Scopes NE, Briggerstaff SM (1973) Progress towards integrated pest control on year round chrysanthemum. Proc 7th Brit Insecticide Fungicide Conf 227
13. Machin MB, Scopes N (1978) Chrysanthemum year-round growing. Blandord, Poole UK
14. Burg SP (1973) Hypobaric storage of cut flowers. HortScience 8:202
15. Waters WE, Conover CA (1967) Chrysanthemum production in Florida. Res Rpt Agr Exp Sta University of Fla, Gainsville
16. Ecke P Jr (1981) Poinsettias. Ohio Flor Assoc Bull 620:1–4
17. Tayama HK (1982) Quality poinsettias-fundamentals 1982. Ohio Flor Assoc Bull 633:1
18. Robertson JL (1981) Poinsettia production costs. Ohio Flor Assoc Bull special Ed 1
19. Ecke P Jr, Matkin OA (1976) The poinsettia mannual. Paul Ecke Poinsettias, Encinitas, California
20. Shanks JB (1980) Poinsettias. In: Larson RA (ed) Introduction to Floriculture. Academic Press, London 301
21. Saltveit ME, Pharr DM, Larson RA (1979) Mechanical stress induces ethylene production and epinasty in poinsettia cultivars. J Am Hortic Sci 104(4):452
22. Saltveit ME, Larsen RA (1981) Reducing epinasty in mechanically stressed poinsettia plants. J Am Soc Hortic Sci 106(2):156
23. Hammer A, Larson RA, Saltveit ME, Reid MS, Sacalis J, Staby G (1982) How to reduce petiole bending of poinsettia plants. Ohio Flor Assoc Bull 626:1–3
24. Larson RA (1982) Poinsettia challenges for the 1980's. Sci Hortic (Amst) 33:113
25. Staby GL, Eisenberg BA, Kelley JW, Bridgen MP, Canningham MS (1980) Leaf petiole epinasty in Poinsettias. HortScience 15(5):635
26. Gilbert DN, Sink KC (1970) The effect of exogenous growth regulators on keeping quality of poinsettia. J Am Soc Hortic Sci 95:984
27. Hammer PA, Krik T (1977) Poinsettia dropy bracts. Focus on Floriculture (Purdue Univ) 5(1):2
28. Sacalis JN (1977) Epinasty and ethylene evolution in petioles of sleeved poinsettia plants. HortScience 12:388
29. Sacalis JN (1978) Ethylene evolution by petioles of sleeved poinsettia plants. HortScience 13(5):594

30. Staby GL, Thompson JF, Kofranek AM (1978) Postharvest characteristics of poinsettias as influenced by handling and storage procedures. J Am Soc Hortic Sci 103:712
31. Staby GL, Kofranek AM (1971) Production conditions as they affect harvest and postharvest characteristics of poinsettia. J Am Soc Hortic Sci 104(1):88
32. Carpenter WJ (1956) The influence of plant hormones on abscission of poinsettia leaves and bracts. Proc Am Soc Hortic Sci 67:539
33. Marousky FJ, Shanks JB (1966) Effects of environmental factors and plant maturity on bract and leaf abscision in *Euphorbia pulcherima* wild. Proc Am Soc Hortic Sci 88:662
34. Gartner JB, McIntyre ML (1957) Effects of day length and temperature on time of flowering of *Euphorbia pulcherima*. Proc Am Soc Hortic Sci 69:492
35. Cathey HM, Campbell LE, Thimijan RW (1981) Strategies for supplemental lighting in greenhouses. How to boost the growth of plants. Ohio Flor Assoc Bull 621:8
36. Kuak DL, Tayama HK, Peterson JC, Fracasso PA (1981) Production of quality 6-inch pinched (one cutting) poinsettia plants from a very early propagation date. Ohio Flor Assoc Bull 620:5, 8
37. Walker JH (1972) Poinsettia and mistletoe toxicity. Society of American Flor (USA)
38. Farnham RB (1976) Results of studies with poinsettias grown in sand. Flor Rev 78 (2007):19
39. Kimmins Kent R (1980) Gloxinias, African Violets and other gesneriads. In: Larson RA (ed) Introduction to Floriculture. Academic Press, London, pp 291
40. Elliot FH (1946) Saintpaulia leafspot and temperature differential. Proc Am Soc Hortic Sci 47:511
41. Went FC (1957) Experimental control of Plant Growth. Chronica Botanica Co Waltham Mass, pp 343
42. Stinson RF, Laurie A (1954) The effect of light intensity on initiation and development of flower buds in *Saintpaulia ionantha*. Proc Am Soc Hortic Sci 75:730
43. Kiplinger DC (1953) Fluorescent lights and saintpaulias. Ohio Flor Assoc Bull 290:2
44. Shanks JB (1960) Some suggestions for planning fertilizer programs. Md Flor 68:4
45. Stinson RF (1952) Saintpaulia tests 1950–51. The effects of light intensity on flower bud initiation and development. The effect of temperature on growth and flower development. Ohio Flor Assoc Bull 271:2
46. Hanchey RH (1945) Effect of fluorescent and natural light on vegetative and reproductive growth of saintpaulia. Proc Am Soc Hortic Sci 66:378
47. Zoerb R (1952) The effect of artificial light on growth and floral development. Ohio Flor Assoc Bull 271:3
48. Boodley JW, Horton FF (1963) Rooting and growth of African Violets under four fluorescent light sources. African Violet Magazine 16(3):103
49. Larson RA (1980) Begonia. In: Larson RA (ed) Introduction to Floriculture. Academic Press, London, 396
50. Horton FF (1949) Christmas begonias in six months. NY State Flower Grow Bull 29:7
51. Post K (1942) Effects of day length and temperature on growth and flowering of some florist crops. Cornell Univ Agric Expt Stn Bull No 787:1
52. White JW (1973) Rieger Elatior begonias: history and European research. Pa Flower Grow Bull 263:1
53. White JW, Guthrie H, Watt B (1973) Rieger Elatior begonias. Research at Pa State Progress Report. IV. Pa Flower Grow Bull 264:8
54. Sandved G (1969) Flowering in *Begonia hiemalis* Fotsch as affected by day length and temperature. Acta Hortic 14:61
55. Runger W, Wehr B (1970) Influence of day length on growth of *Begonia excultorum*. Gartenbauwissenschaft 35:321
56. Molnar JM (1974) Photoperiodic response of *Begonia*×*hiemalis* cv. Rieger. Can J Plant Sci 54:277
57. Mikkelson J (1973) Simplified growing instructions for Rieger Elatior begonias. Pa Flower Grow Bull 263:3
58. Nelson PV, Krauskopf DM, Mignis NC (1977) Visual symptoms of nutrient deficiencies in Rieger elatior begonia (×hiemalis). J Am Soc Hortic Sci 102(1):65

59. Hanisch Ten Cate CH, Berghvef J, Hoorn AMH Van der, Bruinsma J (1973) Hormonal regulation of pedicel abscission in begonia buds. Plant Physiol 33:280
60. Hanisch Ten Cate CH, Bruinsma J (1973a) Abscission of flower bud pedicel of begonia. I. Effects of plant growth regulating substances on abscission with intact plant and with explant. Acta Bot Neer 22:666–674
61. Hanisch Ten Cate CH, Bruinsma J (1973b) Abscission of flower bud pedicel of begonia. II-Interaction and time of sequence of plant regulating substance on abscission of explant. Acta Bot Neer 22:675
62. Hanisch Ten Cate CH, Voogd JJL van der, Bruinsma J (1973c) Abscission of flower bud pedicel of begonia. III-Anatomical pattern of abscission. Acta Bot Neer 22:681
63. Mastalerz JW (1971) Geraniums a manual on culture diseases, insects, economics etc. Pa Flower Grow USA
64. Tayama HK, Poole HA (1976) Extension plants – geranium production – The mother's Day collection. Ohio Flor Assoc Bull 508:6
65. Ball V (1977) Seed geraniums sowing dates. 78 Grow Talks 41(5):1
66. Hammer PA (1980) Other flowering pot plants. In: Larson RA (ed) Introduction to Floriculture. Academic Press, London, 450
67. Craig R, Walker DE (1963) The flowering of *Pelargonium hortorum* Bailey seedlings as affected by solar energy. J Amer Soc Hortic Sci 83:772
68. Carpenter WJ, Rodriguez RC (1971) Early flowering geranium cv. Carefree Scarlet by high intensity supplemental light treatment. HortScience 6(3):206
69. Kanjoinan PS, Tayama HK (1978) Production schedules for seed geraniums. Ohio Flor Assoc Bull 597:1
70. Mastarlerz JW (1967) Geraniums in six weeks. Pa Flower Grow Bull 193:1
71. Ball V, Randalph P (1968) Carefree geraniums spring flowering. Grow Talks 32(4):1
72. Carpenter JW, Carlson WH (1970) The influence of growth regulators and temperature on flowering of seed-propagated geraniums. HortScience 5(3):183
73. White JW (1971) Effects of cycocel, moisture stress and pinching on growth and flowering of F_1 hybrid geraniums (*Pelargonium hortorum* Bailay). J Amer Soc Hortic Sci 95(5):546
74. Love JW (1980) Kalanchoe. In: Larson RA (ed) Introduction to Floriculture. Academic Press, London
75. Irwin JI (1972a) Kalanchoes a new crop. Ohio Flor Assoc Bull 514:1
76. Irwin JI (1972b) Try a "new" crop Kalanchoes. Flor Rev 151(3917):23, 56
77. Anon (1976a) Kalanchoes: background information and production practices for the varieties Mace and Telstar. Yoder Bros, Barbeton, Ohio
78. Anon (1976b) Kalanchoes. Production procedures for year-round production of European Kalanchoe. Yoder Bros, Barteton, Ohio
79. Carlson WH (1975) The culture of *Kalanchoe Blossfeldiana* cultivar. Mace Mich Flor 531: 7, 31
80. Maronsky FJ, Harbaugh BK (1978) Ethylene and temperature interaction. HortScience 13(3):26
81. Mikkelson J (1975) ABC of *Kalanchoe* culture. Ohio Flor Assoc Bull (Short course)
82. Mikkelson J (1977) Focus on floriculture. Focus Floric Purdue Univ 5(1):12
83. Nightingale AE (1970) The influence of succinamic acid. 2,2-dimethyl vs. unpinched plants of *Kalanchoe* hybrid Mace. J Am Soc Hortic Sci 95(3):273
84. Pertuit AJ (1973) The effects of temperature during dark exposure and date of exposure to naturally-occuring day lengths on growth and flowering of *Kalanchoe blossfeldiana* v. Poellnitz. Univ Ga Res Rept No 170
85. Post K (1942) Effects of day length and temperature on growth and flowering of some florist crops. Cornell Agr Exp Sta Bull 787
86. Pertuit AJ (1973) The effects terminal pinching and chemical growth regulation of *Kalanchoe blossfeldiana* Poellnitz. Univ Ga Res Bull No 132
87. Runger W (1967a) About the relationship between formation and development of flowers of various varieties of *Kalanchoe* and the photoperiod and temperature. Gartenbauwissenschaft 32:213

88. Runger W (1967b) Influence of short and long days on juvenile parts of several *Kalanchoe* varieties. Gartenbauwissenschaft 32:399
89. Runger W (1968) On flower formation of *Kalanchoe* Amethyst. Gartenbauwissenschaft 33:345
90. Post K, Horton FF (1950) Give *Kalanchoe* only twenty short days. NY State Flower Grow Bull 57:7
91. Seeley JG (1952) Long days after budding improve *Kalanchoe*. Pa Flower Grow Bull 15:1
92. Yaunis AF (1955) Studies on the photoperiodism of *Kalanchoe blossfeldiana* L. Effects of age on response to short day treatment. Physiol Plant 8:223
93. Harder R (1948) Vegetative and reproductive development of *Kalanchoe blossfeldiana* as influenced by photoperiodism. Sym Soc Exp Biol 2:117
94. Doss RP, Byrne TG, Kretchum TM (1975) *Kalanchoe blossfeldiana* Tetrarulcan suggestions for growing as a pot plant. Flower Nurs Rep Univ Cal 7
95. Batson F (1973) Assembly line Kalanchoe. Flor Rev 152(3943):20
96. Hertough AA De (1980) Bulbous crops. In: Larson RA (ed) Introduction to Floriculture. Academic Press, London
97. De Hertough AA, Roberts AN, Stuart NW, Langhans RW, Linderman RG, Larson RA, Kiplinger DC, Wilkins HF (1971) A guide to terminology of Easter lilies (*Lilium longiflorum* Thunb.) HortScience, 6:121
98. Blanely LT, Roberts AN (1967) Bulb production in Easter lilies. In: Kiplinger DC, Langhans RW (ed) The culture, diseases, insects and economics of Easter lilies. Cornell Univ Press 23
99. Hartley DE (1968) Growth and flowering response of Easter lily (*Lilium longiflorum* Thunb.) to bulb storage. PhD Thesis, Oregon State Univ, Carvallis
100. Lin WC, Roberts AN (1970) Scale function in growth and flowering of *Lilium longiflorum* Thunb. Nellie White. J Am Soc Hortic Sci 95:559
101. Roh SM, Wilkins HF (1974a) Decay and dark reversion of phytochromes in *Lilium longiflorum* Thunb. cv. Nellie White. HortScience, 9:38
102. Hertough AA De, Wilkins HF (1971) Forcing of northwest grown Ace and Nellie White lilies. Flor Rev 149(3857):29. (3858):57, 104
103. Waters WE, Wilkins HF (1967) Influence of intensity, duration and date of light on growth and flowering of precooled Easter lily (*Lilium longiflorum* Thunb. Georgia). Proc Am Soc Hortic Sci 90:433
104. Roh SM, Wilkins HF (1973) The influence and substitution of long days for cold treatments on growth, and flowering of Easter lilies. (*Lilium longiflorum* Thunb. Georgia and Nellie White). Flor Rev 153(3960):19
105. Miller RO, Kiplinger DC (1966) Reversal of vernalization in northwest Easter lilies. Proc Am Soc Hortic Sci 88:646
106. Ling PC, Wilkins HF (1973) The interaction of temperature on photoperiodic response of *Lilium longiflorum* Thunb. cv Nellie White. Flor Rev 153(3965):24
107. Tayama HK, Peterson JC, Lindanist RK, Powel CC, Staby GL (1980) Essentials of pot lily culture. Easter April, 19, Ohio Flor Assoc Bull 612 Oct. 1980
108. De Hertough AA, Carlson WH, Kays S (1967) Controlled temperature forcing of planted lily Bulbs. J Am Soc Hortic Sci 94:433
109. Roh SM, Wilkins HF (1977) The control of flowering in *Lilium longiflorum* Thunb. cv. Nellie White by cyclic or continuous lighting treatments. J Am Soc Hortic Sci 102:247
110. Smith DR, Langhans RW (1962) The influence of photoperiod on the growth and flowering of Easter lily (*Lilium longiflorum* Thunb. var Croft). Proc Am Soc Hortic Sci 80:599
111. Wilkins HF, Roh SM (1970) University of Minnesota's Easter lily Research Report paper No. IX. Lighting lilies at shoot emergence, Night interruption shown to be most effective. Minn State Flor Bull Dec. 10
112. Roh SM, Wilkins HF (1977a) The control flowering in *Lilium longiflorum* Thunb. cv. Nellie White by cyclic lighting. J Am Soc Hortic Sci 102:247
113. Roh SM, Wilkins HF (1977b) The effects of bulb vernalization and shoot photoperiod treatments on growth and flowering of *Lilium longiflorum* Thunb. cv. Nellie White. J Am Soc Hortic Sci 102:229

114. Tizio M, Seeley JG (1976) Nitrogen source, fluoride application and leaf scorch of Ace lilies. Flor Rev 159(4115):43
115. Widmer RE, Mugaas R, Wilkins HF (1976) Lime and phosphorus effects on Easter lilies. Minn State Flor Bull Dec 1
116. Bailey HS (1976) *Hortus third.* McMillian, New York
117. Moe R (1977) Campanula isophylla Moretti culture-cineraria-*Calceolaria* herbeohybrida Voss. Minn State Flor Bull, pp 1–6
118. Runger W (1975) Flower formation in Calceolaria herbeohybrida Voss. Sci Hortic (Amst) 3:45–64
119. Johannson J (1976) The regulation of growth and flowering in *Calceolaria speciosa* Lilj. Acta Hortic 64:239–244
120. Hammer PA (1980) Other flowering pot plants. In: Larson RA (ed) Introduction to Floriculture. Academic Press, London
121. Widered RE (1980) Cyclamen. In: Larson RA (ed) Introduction to Floriculture. Academic Press, London
122. Larson RA (1980) Azalea. In: Larson RA (ed) Introduction to Floriculture. Academic Press, London

Chapter 11 Foliage Plants

11.1 Introduction

The term foliage plants has been used in many ways by several scientists. The diverse nature of the plants included in this group makes it difficult to define the term precisely. Bailey and Bailey [1] used this term to designate plants that are grown primarily for their foliage rather than for flowers. Due to the diverse nature of growth habit and the parts of special interest, there is no single definition of foliage plants which encompasses both these aspects. A recent definition developed by Conover [2] is currently being used in the foliage industry. According to this definition, foliage plants include all the plants grown primarily for their foliage and utilized for interior decoration or interior landscape purposes. The flowers, if present, are inconspicuous and are relatively of secondary importance. Table 11.1 gives the important plant species, including the varieties used for indoor decoration. The importance of foliage plants has increased many folds in the recent past [33, 34].

11.1.1 Quality Requirements

The most important post-production considerations in foliage plants are the beauty of their foliage and the overall form of the plant in relation to container size. Conover [2] suggested other quality requirements such as:

1. Freedom from defects from insect feeding, disease attack, nutrient deficiency, physical and chemical injury.
2. Well-formed, healthy root system that is not pot bound. The largest roots should not come out of the pot holes.
3. The leaves and stems free from chemical residues.

In Europe, foliage condition is given the maximum consideration. An ideal foliage plant must have undergone a minimum period of acclimatization to ensure normal growth of plants when moved to indoor environment.

11.2 Physiology and Biochemistry

11.2.1 Growth and Development

In nature, these plants are grown under tropical conditions of full sunlight, and any change in light intensity is reflected in their ability to retain foliage, and/or

Table 11.1. Some of the important foliage plant grown for indoor use

Name	Family	Common names	Important cultivars
Aglaonema commutatum	Araceae	Aglaonema, Silver King, Silver Queen	Francher white Rajah, Silver Queen, Silver King, Snow Queen
A. modestum			
A. nitidum			
A. treubi			
A. crispum			
A. pseudobracteatum			
Brassaia actinophylla	Araliaceae	Schefflera	–
Cycus revoluta	Cycadaceae	King sago	–
Chamaedorea elegans	Palmaceae	Parlor palm, areca palm	–
Chrysalidocarpus sp.	Palmaceae	Bamboo palm, areca palm	
Dieffenbachia amoena	Araceae	Dumbcane, Dieffenbachia	Bausei, Perfection, Tropic Snow, Rudolph Roehrs, Golden Snow
D. picta			
D. exotica			
Dracaena deremensis	Liliaceae	Dracaena	Janet Craig, Warneckii
D. fragrans			Massangeana
D. godseffiana			Florida Beauty
D. marginata			
D. sanderiana			
Ficus benjamina	Moraceae	Fig, mistletoe fig, rubber tree, weeping fig, etc.	
F. diversifolia			
F. lyrata			
F. pumila			
F. elastica			Cultivars of *Ficus elastica*
F. retusa nitida			Decora Hondura Robusta, Deoscheri
Nephrolepis exaltata	Polypodiaceae	Ferns, Boston fern	Cultivars of *N. exaltate*; Fluffy Ruffles, Florida Ruffle, Rooseveltii
Pteris ensiformis	Polypodiaceae		
Adiantum hispidulum	Polypodiaceae		
Asplenium nidus	Polypodiaceae		
Peperomia sp.	Piperaceae	Peperomia	Emerald Ripple
Philodendron sp.	Araceae	Philodendron	Red Princess, Prince Dubonnet, Emerald King, Emerald Duke
Ptychosperma elegans	Palmae	Solitaire palm	
Scindapsus aureus	Aracease	Devils ivy pothos	Marble Queen
Codiaeum variegatum	Euphorbiaceae	Croton	
Spathiphyllum sp.	Araceae	White anthurium	Mauna Loa, Clere landii
Cacti and succulents	–	Cacti	–
Maranta leuconeura	Marantaceae	Green Prayer plant or Maranta	Kerchoviana
Araucaria sp.	Araucaria+ceae	Norfolk Island pine	–
Chlorophytum comosum	Liliaceae	Spider plant	Vittatum

flowers as well as the dark green leaf colour. Light conditions in the production phase also influence the height of plant, growth rate, area and number of leaves and size of carbohydrate pool in the plant [36–37, 56, 58–59]. Major problems associated with reduction in light intensity during the production phase are undesirable elongation of stems and lower photosynthate production which in turn affect the post-production life of the plant, and induce poor development of leaf variegation [56, 58–59]. In contrast, very high light intensities cause colour fading in some plants (e.g. *Codiaeum* sp.).

The light requirement of foliage plants differs with species and stage of development (Table 11.2). In general, acclimatized plants have lower light intensity requirements under indoor environment (Table 11.3) [2, 39–40, 47–48, 56, 68, 73]. The optimum light intensity for different foliage plants is given in Table 11.2 and 11.3.

Size, shape, and orientation of individual leaf are also influenced by production light. Lower light intensities increase both the number of leaves and total leaf area [32, 37]. Conover [2], working with *Aglaonema* and *Dieffenbachia*, demonstrated that higher light intensities resulted in vertical orientation of leaves as compared to the normal orientation of 60° from the vertical axis under optimum light intensity. Similarly, higher solar radiant flux density caused curving of leaf blade along the mid rib [2]. Production light conditions also affect the leaf anatomy. In *Ficus benjamina*, Johnson et al. [37] demonstrated that sun leaves (0% shade) had heavily cutinized, multiple epidermis characterized by an upper layer of small cells and an additional layer of small subtending cells. The mesophyll had a distinct adaxial palisade layer, a discontinuous irregular abaxial palisade layer and a spongy central and mesophyll layer. The sun leaf had a multiple palisade layer. In contrast, the leaves of shade-grown plants (47% shade) had only limited regions with multiple palisade layer, and had relatively less cutinization in epidermal cells.

Although most of the species included in the foliage plant group are grown primarily for their attractive foliage, some, like Christmas cactus (*Schlumbergera bridgesii*) and *Aphelandra*, are produced for their showy flowers [5, 6]. Flowering of Christmas and Thanksgiving cactus (*S. truncatus*) is controlled by photoperiod and temperature [48, 79]. It produces flowers under short days when the growing temperatures range between 17 and 18 °C, while it is completely inhibited under any photoperiod at 21–24 °C. The plants act like day neutrals at 13 °C. In contrast, Thankgiving cactus initiates flower at 15 °C regardless of photoperiod, while it is completely inhibited at 10 °C [12]. Poole [6, 61] recommended 4 to 6 weeks of short day treatment for successful flowering in *Schlumbergera* sp. Flower development was shown to occur under short or long days. In the case of *Aphelandra*, 8 to 10 h days with 16 to 21 klx light produced profuse flowering [7]. Since most of the plants in this group are native to tropical regions where the length of day and night are relatively constant, they are photoinsensitive in respect to flowering. The vegetative growth in this group is promoted by long days.

In most cases, a night temperature of 18 °C with a day temperature of 24 °C has been found optimum to produce top quality plants [2]. The night tempera-

Table 11.2. Suggested light levels for production of potted acclimatized foliage plants [54]

Botanical name	Light intensity (foot-candles)	Shade cloth (listed as % actual shade)
Aglaonema sp.	2000 – 25 000	80
Aphelandra squarrosa	1000 – 2000	80 winter 90 summer
Araucaria excelsa	5000 – 6000	55
Asparagus sp.	3500 – 4500	63
Brassaia actinophylla	5000 – 6000	55
Calathea sp.	3000 – 3500	73
Chamaedorea erumpens	5000 – 6000	55
Chamaedorea elegans	3000 – 3500	73
Chrysalidocarpus lutescens	5000 – 6000	55
Codiaeum variegatum	7000 – 8000	30
Cordyline terminalis	3500 – 4500	63
Dizygotheca kerchoviana	5000 – 6000	55
Dieffenbachia sp.	3000 – 4500	63 or 73
Dracaena deremensis (cultivars)	3000 – 4500	73 or 63
Dracaena fragrans (cultivars)	3000 – 4500	73 or 63
Dracaena marginata	5000 – 6000	55
Dracaena sp.	3000 – 3500	73
Ficus benjamina	3500 – 4500	63
Ficus elastica (cultivars)	5000 – 8000	55 or 30
Ficus lyrata	5000 – 6000	55
Ficus nidita	3500 – 4500	63
Maranta sp.	2000 – 2500	80
Monstera deliciosa	3500 – 4500	63
(*P. pertussum*)	3000 – 3500	73
Nephrolepis exaltata (cultivars)		
Peperomia sp.	3000 – 3500	73
Philodendron selloum	5000 – 6000	55
Philodendron sp.	3000 – 3500	73
Pilea sp.	2000 – 3500	80 or 73
Sansevieria sp.	3500 – 4500	63
Scindapsus sp.	3500 – 4500	63
Spathiphyllum clevelandii	2000 – 3500	80 or 73
Stromanthe amabilis	2000 – 2500	80
Syngonium podophyllum	3000 – 4500	73 or 63
Zygocactus truncatus	3000 – 3500	73

Table 11.3. Indoor light requirements of potted foliage plants [2]

Low light (0.5 to 0.8 klx)	Medium light (0.8 to 1.6 klx)	High light (1.6 to 3.2 klx)
Aglaonema sp.	*Brassaia, Calathea*	*Aphelandra squarosa*
Chamaedorea sp.	*Chrysalisocarpus* sp.	*Codiaeum variegatum*
Dracaena sp.	*Dieffenbachia maculata*	*Ficus benjamina*
Maranta sp.		*Hoya carnosa*
Peperomia sp.	*Dracaena marginata*	*Nephrolepis exaltata*
Philodendron selloum	*Epipremnum aureum*	
Sansevieria	*Syngonium podophyllum*	

ture of the aerial atmosphere can be reduced to 16 °C if the root zone temperature is maintained at 18–21 °C. In general, lower night temperatures cause accumulation of dry matter, which in turn improves postharvest performance of foliage plants.

Plants like *Aphelandra, Dieffenbachia, Fittonia, Maranta, Epipremnum, Pilea, Aglaonema* and *Scindapsus* are injured at temperatures below 10 °C [8, 62–66]. Chilling injury is commonly associated with collapse of one or more layers of palisade and spongy mesophyll cells forming a horizontal necrotic layer, which eventually cause wilting of the whole plant [66].

Plants of *Ficus benjamina, F. lyrata* and *Codiaeum* have been shown to drop more leaves at very low or very high temperatures. The duration of storage was found to be more important, with longer storage resulting in severe damage and loss of quality [45–46, 72].

Several researchers on various foliage plants have shown the importance of proper nutrition in successful foliage production [4, 15–17, 57, 67–72]. Poole and Conover [72] found that high fertilizer and light levels during production incrased leaf drop following dark storage of *Ficus benjamina.* They further noted that lower fertilizer levels produced chlorotic plants in *Dieffenbachia.* Conover and Poole [42] studied the influence of nitrogen source (NO_3^- , NH_4^+ and urea) on growth of seven foliage genera, *Aphelandra, Aglaonema, Chamaedorea, Dieffenbachia, Nephrolepis, Peperomia* and *Philodendron.* Although there was no clearcut distinction in response of these plants to various nitrogen sources, best *Peperomia* grades were obtained with high nitrate (NO_3) nitrogen source, while top grade *Aglaonema, Chamaedorea* and *Aphelandra* were produced from nitrogen source high in ammonical (NH_4) and urea nitrogen. Uptake of other nutrients in the presence of various nitrogen sources has also been discussed [42].

Most foliage plants are damaged at high salt levels in soils. As the media salt level increases, the water potential of soil approaches that of xylem and reduces water absorption by root [69]. Excessive salts in soil also affect root development, and render the plants more susceptible to soil-borne diseases.

Humidity of aerial atmosphere is yet another environmental factor which needs proper attention from the growers. Most foliage plants grow better at a relative humidity of more than 50% [2].

11.2.2 Senescence of Leaves

The senescence of leaves and flowers has been discussed in detail in Chapter 2.

11.3 Post-Production Losses in Foliage Plants

Post-production losses in quality and quantity of potted foliage plants are related to leaf drop, chlorophyll degradation, uneven or poorly developed variegation in some plants, development of cold water or ring spots on foliage, chilling injury, flower abscission in some, damage of root system, injuries due to mechanical, physiological and pathological causes, and reduction in growth rate of plants. Most of these problems are related to sub-optimal growing conditions in the production area, inadequate acclimatization of plants prior to shipping or lack of training of retailers and consumers on care and handling of these plants.

Cultural factors that affect the qualtiy and post-production life of finished products are light intensity, temperature, humidity, watering and nutrition. Each of these factors has independent effets on growth, but their interaction with each other is more important. Influences of various cultural practices on final plant grade, storability and indoor life have been studied in detail by several investigators [2, 9–11, 36–57, 59, 76].

Acclimatization of plants to indoor environment constitutes a very important phase of production. In this process, the plants adapt themselves to the reduced light and humidity levels that are prevailing in homes. Non-acclimatized plants perform more poorly in storage and indoor situations than acclimatized plants [24, 75].

Conditions during storage and in transit also influence the final grade and lasting quality of foliage plants [26, 28, 43–44, 75–77]. Light intensity and

Table 11.4. Influence of shipping temperature and duration on interior grade and leaf drop in foliage plants [77]

Temperature (°C)	Shipping duration (days)	Dwarf schefflera leaf drop in 12 wks	Weeping fig (*Ficus*)		Areca palm		
			12 wks total leaf drop	Grade[a]	Grade[a]	Chlorophyll content (mg/cm²)	
						At removal	1 wk later
19	7	7.0	9.7	4.2	4.00	0.038	0.048
	14	2.7	22.2	3.5	4.00	0.037	0.037
	21	2.8	34.2	2.7	3.50	0.038	0.040
	28	14.8	42.5	1.7	3.00	0.031	0.048
16	7	2.0	8.2	4.8	4.50	0.040	0.049
	14	5.7	32.0	4.5	3.67	0.039	0.042
	21	2.0	59.7	3.2	3.67	0.046	0.041
	28	11.5	82.2	3.2	3.17	0.043	0.038
13	7	5.3	4.0	4.3	4.00	0.036	0.051
	14	5.5	11.0	3.8	4.00	0.042	0.046
	21	2.2	19.3	3.3	3.83	0.044	0.038
	28	4.8	22.8	2.8	3.33	0.052	0.044
10	7	3.3	9.3	4.7	3.70	0.042	0.052
	14	1.0	19.0	4.3	3.30	0.035	0.041
	21	1.0	37.5	4.0	3.17	0.049	0.046
	28	2.0	57.7	3.7	2.00	0.057	0.048
Controlled plants held in production area		0.7	3.0	5.0	4.33	0.039	0.043

[a] 1 = poor quality, unsaleable, 3 = good quality, 5 = Exellent quality

source, temperature, moisture level, atmospheric pollution and humidity are more important than others [40, 41, 46, 73–77]. The effects of shipping temperatures on various quality factors in some foliage plants is given in Table 11.4. One of the most obvious influences of storing or shipping under sub-optimal conditions is excessive leaf drop, chlorophyll degradation and reduction in plant grade.

The physiological disorders associated with foliage plants are diverse in nature. Most of the physiological disorders are related to mineral deficiencies, as well as toxicities [2], spray injuries, excessive leaf or flower abscission, leaf chlorosis and chilling injuries. Fluoride toxicity is also a major problem in certain foliage plants such as *Dracaena, Calathea, Chlorophytum* and *Maranta*. The symptoms of excessive levels of fluoride in medium include chlorosis or necrosis, which begins at the leaf tip and then proceeds downwards to cover the whole leaf. Some materials like super phosphate, German peat moss, water, perlite and vermiculite, which are rich in fluoride, cause build-up of this element in the soil and produce toxicity symptoms.

Indiscriminate use of pesticides can cause irreversible foliar damage. Several foliage growers use pesticides without following the instructions given on the label.

11.4 Loss Reduction Biotechnology

Use of improved biotechnologies at every stage in the distribution channel ensures delivery of quality products to the consumer. Unlike cut flowers or the potted flowering plants, which are produced for a specific holiday, foliage plants are permanent assets in the home and require constant care and handling. Management of indoor plants is a difficult aspect because of several complicated factors. Another complicating factor in this industry is that the indoor environment is in reality a combination of different environments, involving a range of variables to which these plants are subjected. All these factors make management of foliage plants far more difficult than simply having well-established principles to follow. However, there are certain fundamental considerations which can help to produce and deliver a quality foliage plant; they involve the producer, retailer and consumer, each having a definite role to play. Application of improved biotechnologies, in production, acclimatization, grading, packing, transportation, storage, and handling of finished products at retail and consumer levels can greatly benefit the industry. These biotechnologies can be grouped into:

1. Cultural practices.
2. Acclimatization.
3. Grading, packing and storage.
4. Health management.
5. Use of growth regulators.

11.4.1 Cultural Practices

The optimum light requirement varies with the species (Tables 11.2 and 11.3). Hence, when more than one species are to be grown, it is very important to have arrangements to provide optimum light conditions for each. As low temperatures can cause severe chilling injury to some of the foliage plants, these plants should be produced under optimum aerial (16 to 17 °C night with a day temperature of about 24 °C) [2] and soil temperature (18 to 21 °C).

Nutrition of plants during both the production and post-production period is important. Both the source and method of fertilizer application influence the quality of foliage plants. While in plants like *Diffenbachia* and *Nephrolepis*, plant grade was not influenced by the nitrogen form, and the response was very slight in *Philodendron*, best grades were obtained with nitrogenous fertilizers high in NO_3 in peperomia, and with ammoniacal form of nitrogen and urea in *Aglaonema, Aphelandra* and *Chamaedorea*. Tissue nitrogen levels in these plants were also affected by nitrogen sources [42]. The fertilizer requirements of these indoor plants depend upon prevailing light intensity, with higher requirement at high light intensity [68]. In *Brassaia*, slow release fertilizer was found to be superior to liquid fertilizers under home environment [68].

Poole and Conover [15–16] suggested a fertilizer ratio of 1:1:2 (N:P:K) for foliage plants, and recommended periodical applications on alternate weeks. More detailed information on fertilizer requirements of other plants is also available [2].

11.4.2 Acclimatization

Since most interior foliage plants in their natural habitat are grown at higher light intensities (8000 to 10000 ft.c. during summer), before these sun-grown plants can be used indoors, they will have to be adapted to environmental conditions of interior space, to avoid severe shock leading to complete upset in their physiological functioning. As a result of this shock, some plants lose most of their leaves (*Ficus* and *Schefflera*), while others go into a state of reduced growth and suffer from bad health.

Acclimitization of aerial parts in foliage plants is accomplished in two ways. In the first method, when plants reach a certain height or volume, they are transferred to low light intensity for conditioning and maintained in that situation until they are sold. The period of conditioning varies with species, some requiring only 10 to 12 weeks, while others need up to 1 year. By the second method, the plants are grown at specific shade levels from the beginning and are sold whenever a desired size is attained. The latter method takes a longer time to produce marketable-sized plants. This system is best suited for plants which are small and are able to grow at even relatively low light intensities. Some of the cultural conditions, such as production light and fertilization, influence the ability of the plants to acclimatize.

Ideally, the acclimatization process involves both light acclimatization and soil or root-mass acclimatization. Light acclimatization brings about changes in leaf anatomy by the reorganization of cells in epidermal and palisade layers to adjust themselves to low light intensity. When the plants are placed in a low light environment, the plants, depending upon the species, may lose some of their leaves. But if the other factors are maintained at optimum levels, these plants will develop leaves that are capable of conducting photosynthesis under low light environments. As the chloroplasts become reoriented in the leaf cells, the foliage looks much darker. In plants which do not shed leaves under low light conditions, the greening-up process is the major part of acclimatization. In contrast, in plants like *Ficus* and *Schefflera*, the acclimatization process results in thicker leaves. They generally take a longer time to adapt themselves to low light situations. Newly developed leaves in these plants are better prepared for adverse conditions during storage or indoor environment. It is important to realize that the acclimatization process is nonreversible in nature and exposure to light intensities around 2000–3000 ft. c. can severely damage the plant by killing the chloroplast [18].

There are definite specifications for acclimatized foliage plants. These specifications are based on species and container size [18]. Acclimatized plants that fulfil all the established specifications will continue to grow and maintain top quality under indoor environment and pose fewer problems in their maintenance. Minimum specifications for indoor foliage plants are available elsewhere [78].

In general, sun-grown plants have a more active root system and a different shoot-to-root ratio than when they are placed in an indoor situation. This is mainly based on need, as under high light intensity the foliage is actively involved in photosynthesis and transpiration with the result that the root system has to balance foliage growth by producing new roots. At low light conditions, both photosynthesis and transpiration are reduced, thereby resulting in a disturbed balance between the aerial and underground portion. Therefore, an efficient acclimatization process must involve both foliage and root acclimatization. The main objective of root acclimatization is to establish a top-to-root ratio suited to indoor environment, particularly in plants which lose their leaves. This is accomplished by regulating soil moisture levels. Studies by Conover [2] have clearly indicated that with proper regulation of shade levels and watering, it is possible to establish a new root : shoot equilibrium. Conover [2] listed the following steps for achieving the best results with acclimatization in foliage plants:

1. Leach the potting medium to remove excess fertilizers and wait until 2 to 3 months to begin indoor fertilization. Plants indoors need about one-tenth the nutrients of their counterparts grown under high light intensity.
2. Maintain the plants in high humidity environment for first 1 to 2 weeks.
3. Maintain the plants under at least 1.6 klx of light during first 1 to 2 weeks.
4. Regulate soil water in pot to avoid root damage.

11.4.3 Grading, Packing and Storage

Grading. The grading system and the grade specifications presently followed in the foliage industry are highly variable and mainly based on the experience of the growers. Most grading systems, whether developed by the grower or otherwise, are based on the quality of the foliage (colour, size and freedom from injuries), height and width of plant, number of shoots per pot, and the overall appearance of the plant in relation to the pot size [18]. The European foliage industry follows a more stringent system of grading than the United States. In general, a top quality plant must be well developed to occupy the whole space in the container, with foliage representative of the species, or cultivar, and free from damage from diseases, insects, physiological disorders, chemical residues and physical injuries. A good, healthy, well-developed root system is an indication of a top quality plant. Some of the plants, like *Araucaria, Dracaena* and *Brassaia,* are sold according to size or age.

Packing. Proper packing technique can minimize the damage occurring from desiccation and darkness in transit. The actual packing method depends upon the species and size of plant, the marketing outlet and also mode of transportation [25–27]. There are two main kinds of packing systems followed in the foliage industry. In the first method, the plants are enclosed in paper, plastic or fibre-glass bags before placing them in standard packing boxes. Usually plants in smaller containers do not need sleeves, but large-sized plants necessarily require sleeves. Some growers use dividers to support individual plants in boxes and to prevent displacement in transit. It is very important to make sure that the sleeves, as well as the packing box, be made from materials that do not absorb moisture from plants. These boxes are usually coated with wax to prevent desiccation during transport. Some growers use waxed trays before placing them in boxes. Hanging baskets and large-sized plants are boxed individually by providing sufficient protection to avoid friction between plant and packaging material. It is important to make sure that the plants have an adequate supply of water to meet their requirements during transportation or holding.

The most common packaging material used is cardboard boxes of different sizes. Research by Harbaugh et al. [30] revealed build-up of high levels of ethylene, increased disease problems, and poor response of plants in adapting to low humidity conditions in indoor environment. Further investigations are needed to overcome these problems and to make see-through boxes a reality and practicable.

The second method is followed mostly by progressive growers of foliage plants, who operate their own truck lines to supply the plant materials. These trucks have built-in racks to accommodate individual containers of different sizes. This method thus saves a lot of time in packing as well as money on packing boxes. However, these plants are often wrapped in paper or pastic sleeves. In this method, there is very little chance of desiccation injury or damage due to fluctuations in the shipping environment. The shipping environ-

Table 11.5. Storage requirements for important indoor foliage plants [31]

Species	Harvest Factors	Storage temperature (°C)	Storage relative humidity (%)	Storage light (ft.c.)	Remarks
Aglaonema sp.	1. The root system should be well developed at the time of harvest 2. Soluble salts below 1500 ppm	18.2±2.8	75±10	50	1. Under normal conditions can be stored up to 10 days, after 14 days leaf drop will occur 2. Can be stored indefinately with artificial lighting
Brassaia sp.	1. Root system at the time of harvest must be extensive and healthy 2. Soluble salts below 1000 ppm	15.4±2.8	75±10	150	1. Water stress causes severe leaf drop 2. Under normal conditions it can be stored upto 7 days
Chamaedorea sp.	1. Should be well established 2. Extensive healthy root system 3. Soluble salts below 1000 ppm	15.4±2.8	65±10	150	1. Under normal conditions 10 to 14 days 2. Under light indefinitely
Dieffenbachia sp.	1. Well established root system 2. Soluble salts below 1500 ppm	18.2±2.8	75±10		1. Can be seriously damaged at temperature below 12.7 °C 2. Under normal conditions can be stored upto 5 to 7 days 3. Temperature should be above 15.4 °C

ment can be modified according to the requirements of the consignment. It enables the growers to ship their plants over a long distance in a relatively short time and at low cost.

Foliage plants are transported through trucks, air or sea. The type of containers used for packing and the conditions in transit will largely determine the extent of losses during shipping and also the post-transportation life of these plants [19–23, 25–28, 41–43, 45, 72, 77).

Table 11.5 (continued)

Species	Harvest Factors	Storage temperature (°C)	Storage relative humidity (%)	Storage light (ft.c.)	Remarks
Dracaena sp.	1. Well-established root system 2. Soluble salts below 1200 ppm	18.2 ± 2.8	75 ± 10		Normal storage life 7 days
Ficus sp.	1. Well established root system 2. Soluble salts below 1200 ppm	15.4 ± 2.8	65 ± 20		1. Normal storage life 7 days 2. Ethylene levels should be below 2 ppm
Nephrolepis sp.	1. Plants should be well established before sale 2. Soluble salt below 1000 ppm	18.2 ± 2.8	75 ± 10		Normal storage life 7 days
Peperomia sp.	1. Well-established root system 2. Soluble salts below 1000 ppm	18.2 ± 2.8	75 ± 10	100	1. Normal storage life 7 days
Philodendron sp.	1. Well-developed root system 2. Soluble salt level below 1500 ppm	18.2 ± 2.8	75 ± 10	–	Can be stored up to 7 to 10 days
Scindapsus aureas	1. Extensive and healthy root system 2. Soluble salts below 1500 ppm	18.2 ± 2.8	75 ± 10	–	1. Normal storage life 7 days 2. Ethylene levels above 2 ppm cause injuries

Storage. The studies on storage of foliage plants are limited as compared to cut flower crops [31, 40, 41, 76, 77). Storage of foliage plants has become important due to the involvement of newer marketing outlets, such as mass markets. Plants in these situations have to be stored for a relatively longer period of time without any adverse effects on their grade and lasting quality. Furthermore, more and more foliage plants are being shipped by trucks or by sea freight, involving an extended period of enclosure of plants in shipping environment. Concise information on the storage of foliage plants is presented in Table 11.5.

11.4.4 Maintaining the Health of Foliage Plants in the Home

As indicated earlier, maintaining the health of foliage plants in an interior environment is complicated due to several factors, such as the wide range of plant species included in this group, components of indoor environment (light, temperature, humidity, and nutrition) and involvement of different kinds of people in the delivery of these plants.

11.4.5 Control of Pests and Diseases

Many people believe that if the plant is free of diseases and pests at the time of purchase, the home owner will have very little disease or insect problems in the indoor environment [13, 14]. However, some of the root problems are common under overwatered or poorly aerated soil, so proper plant protection measures are necessary in the greenhouse production to prevent these problems.

11.4.6 Use of Growth Regulators

Yonemura [79] reported that benzyl adenine (BA) promoted the number of buds under inductive conditions and also increased the number of phyllodes in *Schlumbergera*. Similar promotion of flowering was obtained with 1000 ppm of GA_3 in *Spathiphyllum* Mauna Loa. However, GA-treated plants had distorted flowers [80]. Foliar application of silver thiosulphate reduced flower and bud abscission of zygocactus plants exposed to 0.5 m/per litre of ethylene [79].

References

1. Bailey LH, Bailey EZ (1976) Hortus Third. McMillan, New York
2. Conover CA (1980) Foliage plants. In: Larson RA (ed) Introduction to Floriculture. Academic Press, London 555
3. Graf AB (1970) Exotica B. Roehrs Rutherford, New Jersey USA
4. Rodriguez SJ, Cibes H (1977) Effects of five levels of nitrogen at six shade intensities on growth and leaf-nutrient composition of *Dracaena deremensis* Warneckii Engler. J Agric Univ P-R 61:305
5. Christensen OV (1969) The influence of low temperature on flowering of *Aphelandra squarrosa*. Tridsskr Planteavl 73:351
6. Poole RT (1971) Flowering of Christmas Cactus as influenced by nyctoperiod regimes. Proc Fla Sta Hortic Soc 84:410
7. Kerbo R, Peyane RN (1976) Reducing flowering in *Aphelandra squarrosa* Nees with high pressure sodium lighting. HortScience, 11:368
8. McWilliams EL, Smith CW (1978) Chilling injury in *Scindapsus pictus, Aphelandra squarrosa* and *Maranta leuconeura*. HortScience, 13:179
9. Conover CA, Roole RT (1977a) Effects of cultural practices on acclimatization of *Ficus benjamina*. J Am Soc Hortic Sci 102:529
10. Vlahos J, Boodley JW (1974) Acclimatization of *Brassaia actinophylla* and *Ficus nitida* for interior environmental conditions. Flor Rev 154:18
11. Collard RM, Joiner JN, Conover CA, McConnell DB (1977) Influence of shade and fertilizer on light compensation point of *Ficus benjamina*. J Am Soc Hortic Sci 102:477
12. Fanteno WC, McWilliams EL (1978) Light compensation points and acclimation of our tropical foliage plants. J Am Soc Hortic Sci 103:52
13. Powel CC (1982) Managing the health of indoor plants. Ohio Flor Assoc Bul 633:4
14. Hamlen RA, Short DE, Henley RW (1975) Detect and control insects and pests on tropical foliage plants. Flor Rev 9:72

15. Poole RT, Conover CA (1977) Nitrogen and potassium fertilization of *Aglaonema commutatum* Schott. cvs Fransher and Pseudobracteatum. HortScience 12:570
16. Poole RT, Conover CA (1978) Fertilization on Maidenhair fern, *Adiantum radianum* K. Presl. HortScience 13:176
17. Poole RT, Conover CA (1976) Chemical composition of good quality tropical foliage plants. Proc Fla Sta Hortic Sci 89:307
18. Gains RL (1978) Guidelines for foliage plant specifications for interior use. Fla Foliage Assoc Apopka
19. Akamine EK, Goo T (1977) Test shipment of Hawaii-grown ornamentals by surface transportation. Hawaii Agric Exp Stn Res Rept 52:1
20. Akamine EK, Goo T (1978) Test shipment of ornamentals by surface transportation. Univ Hawaii Coop Extn Ser Misc Pub 172:44
21. Angus DE, Bielorai H (1965) Transpiration reduction by surface films. Aust J Agric Res 16:107
22. Anon (1979) Shipping container keeps plants in top condition. Flower Marketing 86(34):8B, 11B
23. Anon (1980) Requirements on containers, packing material, methods of packing and labeling of export plant nursery products Part A Report on Nursery Products – packaging and standards. Hawaii Dept Agric pp 1
24. Conover CA, Poole RT (1978) Influence of ligth intensity on fertilizer requirements of acclimatized foliage plants maintained under an interior environment. HortScience 13(3):363
25. Conover CA, Poole RT (1982) Results of the Florida test shipment. Fla Nurserym 29(4):28
26. Hugues M (1978) Controlled shipping ensures maintenance of quality foliage. Flor Rev 161 (4183):24
27. Mitchel H (1982) Industry moves from cut flowers to potted plants. Flor Directions 7(9):1
28. Saltveit MF (1980) An inexpensive chemical scrubber for oxidizing volatile organic contaminants in gases and storage room atmospheres. HortScience 15(6):759
29. Wilson SL, Peterson JC (1982) A comparative index of water use for flowering and foliage plants. HortScience 17(3):519
30. Harbaugh BK, Wilfret GJ, Engelhard AW, Water WE, Marousky FJ (1976) Evaluation of 40 ornamental plants for a mass marketing system utilizing sealed polythene package. Proc Fla Sta Hortic Soc 89:320
31. Staby GL, Robertson JL, Kiplinger DC, Conover CA (1978) Chain of life. Ohio Flor Assoc Bul Ohio USA
32. Cathey HM, Compbell RW (1978) Zero-base budgeting for lighting plants. Foliage Dig 1:10
33. Anon (1982) Floricultural crops, production, area and sales. Statistical reporting service USDA Washington DC
34. Anon (1971) Production, area and sales of floricultural crops. Statistical reporting Service USDA Washington DC
35. Harborough EF (1977) Fresh foliage plant summary. Fed State Mark News Serv Orlands Florida
36. Collins PC, Blessington TM (1981) Influence of production light and ancymidol on foliage plants. HortScience 16:215
37. Johnson CR, McConnell DB, Joiner JN (1982) The influence of ethaphon and light intensity on growth and acclimatization of *Ficus benjamina*. HortScience 17(4):614
38. Ben-Jaacov I, Poole RT, Conover CA (1982) Effects of long-term dark storage on quality of *Shefflera*. HortScience 17(3):347
39. Garvey EJ, Blessington TM (1981) Effects of production solar radiant flux density and duration of simulated shipping on the interior keeping quality of *Ficus elastica* Roxb e.g. Hornem Decora. HortScience 16:290
40. Brasswell JH, Blessington TM (1981) Effects of production shade levels and subsequent dark storage on interior quality retention' of *Brassaia species*. HortScience 16(3):290
41. Buck TL, Blessington TM (1982) Postharvest effects of temperatures during simulated transit on quality factors of two *Ficus* species. HortScience 17(5):817
42. Conover CA, Poole RT (1982) Influence of nitrogen source on growth and tissue nutrient content of foliage plants. HortScience 17(3):90

43. Auer CA, McConnell DB (1982) Effects of transit stress on Begonia Medora and dwarf *Schefflera*. HortScience 17(3):100
44. Collins PC, Blessington TM (1982) Postharvest effects of dark storage, light duration and light source on keeping quality of *Ficus benjamina*. HortScience 17(3):100
45. Poole RT, Conover CA (1982) Influence of shipping temperatures on long-term dark storage and interior longevity of foliage plants. HortScience 17(3):100
46. Bequuette BL, Blessington TM (1982) Postharvest effects of light duration on growth and quality of *Codiaeum variegatum* Pictum cultivars as affected by production light levels. HortScience 17(3):100
47. Mbah BN, McWilliam EL (1982) Carbon balance of *Peperomia obtusifolia* plants during low light acclimatization. HortScience 17(3):100
48. Heins RD, Amertage AM, Carlson WH (1981) Influence of temperature, water stress, and BA on vegetative and reproductive growth of *Schlumbergia truncata*. HortScience 16(5):679
49. Conover CA, Poole RT (1975) Acclimatization of tropical tress for interior use. HortScience 10:600
50. Peterson JC, Sacalis JN, Durkin DJ (1980a) Promotion of leaf abscission in intact *Ficus benjamina* by exposure to water stress. J Am Soc Hortic Sci 105(6):788
51. Peterson JC, Sacalis JN, Durkin DJ (1980b) Alterations in abscisic acid content of *Ficus benjamina* leaves resulting from exposure to water stress and its relationship to leaf abscission. J Am Soc Hortic Sci 105(6):793
52. Peterson JC, Sacalis JN, Durkin DJ (1981) *Ficus benjaminia* – avoid water stress to prevent leaf sheading. Flor Rev 167(4335):10, 11, 37, 38
53. Conover CA, Poole RT (1975) Influence of shade and fertilizer levels on production and acclimatization of *Dracaena marginata*. Proc Fla Sta Hortic Soc 88:606
54. Conover CA, Poole RT, Henley RW (1975) Growing acclimatized foliage plants. Fla Foliage Grower 129
55. Milks RR, Joiner JN, Garard LA, Conover CA, Tjia B (1979) Influence acclimatization on carbohydrate production and translocation of *Ficus benjamina* L. J Am Soc Hortic Sci 104:410
56. Johnson CR, Nell TA, Rosenboum SE, Lauritis JA (1982) Influence of light intensity and drought stress on *Ficus benjamina* L. J Am Soc Hortic Sci 107:252
57. Conover CA, Poole RT (1976) Light and fertilizer recommendations on production of foliage stock plants and acclimatized potted plants. Univ Fla Agric Res Rep RH-76-6·Sp
58. Johnson CR, Nell TA, Joiner JN, Kranz JK (1979) Effects of light intensity and potassium on leaf stomatal activity of *Ficus benjamina* L. HortScience 14:227
59. Johnson CR, Kranz JK, Joiner JN, Conover CA (1979) Light compensation points and leaf distribution of *Ficus benjamina* as affected by light intensity and nitrogen-potassium nutrition. J Am Soc Hortic Sci 104(3):335
60. Fails BS, Lewis AJ, Borden JA (1982) Light acclimatization potential of *Ficus benjamina*. J Am Soc Hortic Sci 107(5):762
61. Poole RT (1973) Flowering of christmas cactus during summer. HortScience 8:186
62. Fooshee WC, McConnell DB (1980) The effect of chilling on subsequent rooting of *Aglaonema* Silver Queen tip cuttings. Proc Fla Sta Hortic Soc 93:212
63. Marlet RB (1974) Chilling injury in *Sanseveria*. HortScience 9:539
64. McConnell DB, Ingram DI, Groga-Brada C, Sheehan TJ (1970) Effects of chilling on propagation of *Dieffenbachia* Exotic Perfection. Proc Trop Reg Am Soc Hortic Sci 26:285
65. McWilliams EL, Smith CW (1978) Chilling injury in *Scindapsus pictus, Aphelandra squarrosa* and *Maranta leuconeura*. HortScience 13:179
66. McConnell MB, Ingram DL, Groga-Blada C, Sheehan TJ (1982) Chilling injury in silvernerve plant. HortScience 17(5):819
67. Conover CA, Poole RT (1983) Influence of ligth and fertilizer during production on simulated shipping of *Ficus benjamina*. HortScience 18(4):606
68. Conover CA, Poole RT (1981) Influence of light and fertilizer levels and fertilizer sources on foliage plants maintained under interior environment for one year. J Am Soc Hortic Sci 106:571

69. Nolan SL, Ashe TH, Lindstrom RS, Martens DC (1982) Effects of sodium chloride levels on four foliage plants grown at two light levels. HortScience 17(5):815
70. Jonier JN, Conover CA (1981) Nutrition and fertilization. In: Joiner JN (ed). Foliage plant Production Prentice-Hall pp 256–263
71. Conover CA, Poole RT (1977b) Influence of irrigation method and fertilizer source and level on growth of four foliage plants. Proc Fla Sta Hortic Soc 90:312
72. Poole RT, Conover CA (1979) Influence of shade and nutrition during production and dark storage simulating shipment on subsequent quality and cholorphyll content of foliage plants. HortScience 14:617
73. Peterson NC, Blessington TM (1981) Postharvest effects of dark storage and light source on keeping quality of *Ficus benjamina* L. HortScience 16:681
74. Bellinger P, Griffith S (1978) Florida Foliage Buyer's Guide. Fla Foliage Buyer's Guide, Apopka USA
75. Conover CA, Poole RT (1975) Acclimatization of tropical foliage plants. Univ Fla Apopka Res Rep, RH-75-3
76. Batson DB, Blessington TM (1983) Influence of postharvest light levels on long term effects of dark storage on the postharvest keeping quality of *Schefflera arboricola*. HortScience 18(2):82
77. Poole RT, Conover CA (1983) Influence of simulated shipping environments on foliage plant quality. HortScience 18(2):191
78. Gaines RL (1977) Interior Plantscaping. Architectural Record. McGrow-Hill New York p 182
79. Yonemura K (1979) Studies on the control of flowering Christmas cactus. Sp Bull Aichiken Agr Res Cen Nagakute Aichi Japan
80. Henny RJ (1981) Promotion of flowering in *Spathiphyllum* Mauna Loa with gibberellic acid. HortScience 16(4):554

Subject Index

MIX
Papier aus verantwortungsvollen Quellen
Paper from responsible sources
FSC® C105338

If you have any concerns about our products,
you can contact us on
ProductSafety@springernature.com

In case Publisher is established outside the EU,
the EU authorized representative is:
Springer Nature Customer Service Center GmbH
Europaplatz 3, 69115 Heidelberg, Germany

Printed by Libri Plureos GmbH
in Hamburg, Germany